Implantable Electronic Medical Devices

Implantable Electronic Medical Devices

Dr Dennis Fitzpatrick PhD CEng BEng(Hons)

MIET MIEEE FHEA

AMSTERDAM • BOSTON • HEIDELBERG • LONDON
NEW YORK • OXFORD • PARIS • SAN DIEGO
SAN FRANCISCO • SINGAPORE • SYDNEY • TOKYO

Academic Press is an imprint of Elsevier

Academic Press is an imprint of Elsevier
525 B Street, Suite 1800, San Diego, CA 92101-4495, USA
225 Wyman Street, Waltham, MA 02451, USA
The Boulevard, Langford Lane, Kidlington, Oxford OX5 1GB, UK

Notices
Knowledge and best practice in this field are constantly changing. As new research and experience
broaden our understanding, changes in research methods, professional practices, or medical treatment
may become necessary.

Practitioners and researchers must always rely on their own experience and knowledge in evaluating
and using any information, methods, compounds, or experiments described herein. In using such information
or methods they should be mindful of their own safety and the safety of others, including parties for whom
they have a professional responsibility.

To the fullest extent of the law, neither the Publisher nor the authors, contributors, or editors, assume
any liability for any injury and/or damage to persons or property as a matter of products liability, negligence
or otherwise, or from any use or operation of any methods, products, instructions, or ideas contained
in the material herein.

ISBN: 978-0-12-416556-4

British Library Cataloguing-in-Publication Data
A catalogue record for this book is available from the British Library

Library of Congress Cataloging-in-Publication Data
A catalog record for this book is available from the Library of Congress

For Information on all Academic Press publications
visit our website at http://store.elsevier.com/

Typeset by MPS Limited, Chennai, India
www.adi-mps.com

Printed and bound in the United States

Contents

Preface

Implantable systems in the human body are now becoming more widely acceptable and available since the development of pacemakers and other implantable electronic medical devices (IEMDs) such as hearing aids and glucose sensors. With a greater life expectancy and an increasing demand for medical healthcare, there is a greater demand on technology and biomedical engineers to develop implantable systems for a wide variety of medical diagnostics, treatments and therapies. It is fortunate that technology has advanced to complement the realization of IEMDs in terms of miniaturization, complexity, biomaterials and defined standards.

With the new IEMDs comes the introduction of new regulatory standards to bring together national standards from different countries under one internationally agreed standard for the safe design and implementation of IEMDs. Manufacturers will have to comply and show compliance with the new regulatory standard by displaying the new certification marks.

This book collectively groups medical devices with similar functionality into separate chapters. It is the intention of this book to provide a background on the application of medical devices, an introduction to the latest techniques used and examples of existing medical devices. Subsequently, the book can be used as a guide to the design of medical devices and also as a reference for existing medical devices. The book is aimed at those involved in or who have an interest in the research and design of IEMDs. The healthcare industry is vast and includes electronic engineers, bioengineers, biomedical engineers, clinical engineers, clinical scientists, medical practitioners, surgeons and students alike.

Every effort has been made to provide an accurate description of the IEMDs featured in this book. Consequently, I am very grateful to the respective representatives from the medical device manufacturers for their invaluable feedback in order to ensure an accurate representation of their implantable devices. My thanks also go to Laurel Brumant for the anatomical illustrations and to Dr. David Chappell from the University of West London for proofreading and reviewing the chapters in the book. I also thank Barry Nevison for his support on cochlear implants and Tina Lee for her information research. A big thank you also goes to Cari, Fiona and Naomi at Elsevier for their perseverance in chasing up device manufacturers.

DISCLAIMER

Although every effort has been made by the author to ensure an accurate description of the implantable electronic devices featured in this book, the author cannot be held responsible for any inaccurate representation of the featured IEMDs.

Chapter

1

Retinal Implants

1.1 INTRODUCTION

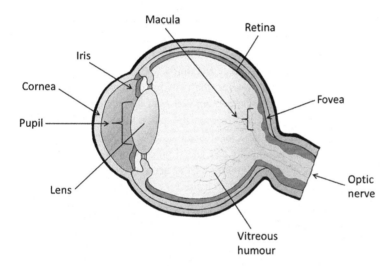

Macula

Retina

Iris

Cornea

Fovea

Pupil

Optic nerve

Lens

Vitreous humour

■ **FIGURE 1.1** Structure of the eye.

Figure 1.1 shows the main anatomical features of the eye. In normal sight, light enters the eye through the pupil and is focused onto the retina at the back of the eye, stimulating photocells that translate the light into electrical signals. These electrical signals travel down the optic nerve to the visual centers in the brain where they are decoded and perceived as images. Progressive diseases of the eye that result in partial or total loss of vision include glaucoma, retinitis pigmentosa, and macular degeneration.

Glaucoma results from an increase in the internal pressure of the eye, the effects of which are irreversible, eventually leading to loss of sight. However, if detected early, the onset of the disease can be managed with medical treatment or laser surgery. Measuring the intraocular pressure of the eye can help in detecting the early stages of the disease (see Chapter 2).

Implantable Electronic Medical Devices. DOI: http://dx.doi.org/10.1016/B978-0-12-416556-4.00001-2

Retinitis pigmentosa is a genetic disorder resulting in the degeneration of the photoreceptor cells in the retina, leading to partial or complete loss of sight. Currently there is no cure, although gene therapy in which a virus is used to deliver sight-restoring therapeutic genes to the photoreceptors at the back of the eye may offer an alternative form of treatment in the future.

Age-related macular degeneration (AMD) is another disease of the retina, but it only affects a small area of the retina known as the macula which contains a small population of cone-type photoreceptor cells that are more responsive to bright light levels required for reading and viewing objects close up and in greater detail. The onset of AMD occurs in the later stages of life and only leads to a partial degeneration of sight.

Retinal implants are used to help people with degenerative retinal diseases such as retinitis pigmentosa and AMD where the optic nerve and the visual centers in the brain are still functioning but the patient has lost light or sight perception due to degeneration of the outer layer of the retinal photoreceptor cells. However, the cells in the inner retinal layer are relatively intact compared to the outer cells and it is the inner cells which form a neuronal ganglion interface to the optical nerve. Retinal implants will not benefit people who have been blind from birth because their optical visual neuronal circuits and visual processing centers in the brain have not been developed or conditioned to perceive vision.

1.2 THE RETINA

Light entering the eye through the lens is focused onto the retina which consists of a thin layer of transparent neural tissue located at the back of the eye. Near the center of the retina is a region known as the macula which has a high concentration of neural cells responsible for seeing detailed colors and represents the center of vision. At the center of the macula is a small depression or dimple known as the fovea which represents the absolute center of vision and highest color resolution attainable, providing the clearest and sharpest images. Subsequently, the eye continuously moves (saccades) such that the lens focuses images of interest onto the fovea for the highest image of color resolution.

The retina is made up of three main functional neural cell layers: photoreceptor cells, bipolar cells, and ganglion cells. Interspersed between the layers are the horizontal and amacrine neural cells as shown in Figure 1.2. The photoreceptor cells at the back of the retina transduce photon light energy into graded neural signals which are transmitted and processed via the bipolar and ganglion cell layers. It is the axons of the

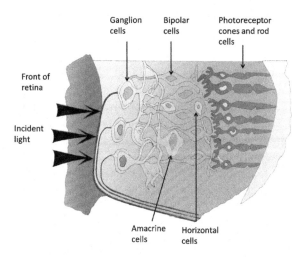

■ **FIGURE 1.2** Structure of the retinal layers.

ganglion cells which together collectively form the optic nerve which leads to the visual processing centers in the brain.

1.3 PHOTORECEPTOR CELLS

There are two types of photoreceptor cells: rods, which have the ability to detect color but are sensitive to low light levels (scotopic vision), and cones, which in bright light are sensitive to colors (photopic vision) in the visible spectrum. The rods and cones are made up of four segments

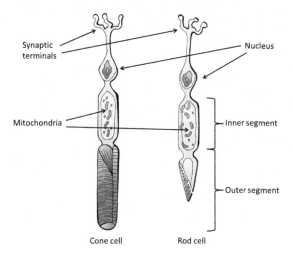

■ **FIGURE 1.3** Photoreceptor cone and rod cells.

(Figure 1.3): the outer segment, inner segment, cell body (nucleus), and synaptic terminals.

The outer segment in rods and cones consists of the outer membrane folding in on itself and stacking up to form disks. In the case of rods, the infolded membranes become detached and the disks float inside the outer segment. Located on the disks are light-sensitive pigment proteins, rhodopsin in rods, and iodopsin in cones. The inner segment contains mitochondria which provide the energy required for chemical reactions and the cell body which contains the cell nucleus and other cell organelles essential to maintain cell functionality. The synaptic terminals provide for the transmission of glutamate neurotransmitters between neural cell synaptic bodies.

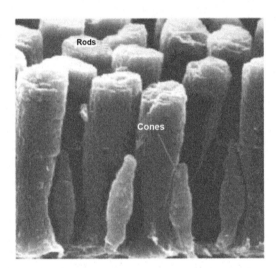

■ FIGURE 1.4 Structure differences between rods and cones. *(http://www.ic.ucsc.edu/ ~ bruceb/psyc123/ Vision123.html.pdf.)*

In rods, the outer segment is cylindrical, whereas for cones, the outer segment is conical in shape (Figure 1.4). Typical outside diameters for the inner and outer segments are 2 μm for rods and 6 μm for cones. The rods also contain a greater number of light-sensitive disks in the outer segment compared to cones, resulting in a greater sensitivity to light. There are typically 120 million rods compared to 6 million cones in the retina.

In rods, all the disks contain the same light-sensitive pigment, rhodopsin, which exhibits a peak absorption of light energy at a wavelength of 500 nm which lies within the blue-green region of the visual light spectrum. In cones, the light-sensitive iodopsin pigment occurs in three varieties due to differences in their amino acid sequence, each with different peak absorption wavelengths in the red (560 nm), blue (420 nm), and green (530 nm) regions of the visible light spectrum, respectively.

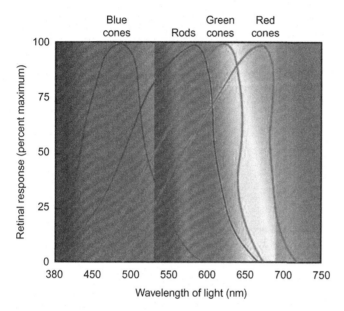

■ **FIGURE 1.5** Electromagnetic spectrum of the human eye.

Although each cone contains three different opsin pigment types, there are three different types of defined cones: short-wave (blue light), medium-wave (green light), and long-wave (red light), each with a predominant opsin variety in the cone. The superimposition of the light absorption response of each opsin pigment will result in a peak response around the area of the defined cone color type. For example, the peak response of a long-wave cone will be shifted due to the superimposition of the individual blue and green opsin spectrum absorption responses, toward the yellow-green region of the visible spectrum as shown in Figure 1.5.

Figure 1.6 shows a rod photoreceptor cell with sodium- and potassium-specific ion channels in the outer membrane. In the absence of light, there will be a continuous flow of positively charged sodium ions into the cell and potassium ions out of the cell, collectively known as the "dark current." This dynamic arrangement gives the photoreceptor cell a resting potential of approximately -30 to -40 mV. Neurotransmitters (glutamate) are also released from the synaptic terminals of the photoreceptor cell. When light photons strike the visual pigments in the disks, a

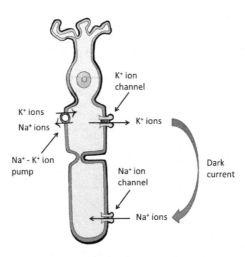

■ **FIGURE 1.6** Induced ionic currents in photoreceptor cell.

series of chemical reactions involving enzyme activity causes the cell to hyperpolarize and reduce the release of synaptic neurotransmitters.

1.4 **BIPOLAR AND GANGLION CELLS**

As shown in Figure 1.2, the bipolar and ganglion cell layers are interlaced with two other cell types, the horizontal and amacrine cells. The neural signals from the photoreceptor cells interface with the bipolar cells directly or indirectly via the horizontal cells, which in turn interface with other bipolar cells or other adjacent horizontal cells. Similarly, the bipolar cells interface with the ganglion cells directly or indirectly via the amacrine cells, which in turn interface with other ganglion cells and other adjacent amacrine cells.

There are two types of bipolar cells, both of which receive the glutamate neurotransmitter, but the ON-center bipolar cells will depolarize, whereas the OFF-center bipolar cells will hyperpolarize. This arrangement helps provide a spatial processing of the visual input derived from the photoreceptor cells. The bipolar cells provide one of many sensory inputs to the ganglion cells which are thought to be involved with temporal aspects of color vision being sensitive to speed of movement. The output synapses of the ganglion cells form the optic nerve which transmits the neural image data to the visual cortex in the brain for decoding into perceived images. The ganglion cells also contain the photopigment melanopsin which is involved in the pupillary light reflex mechanism where the pupil constricts when the retina is exposed to bright light.

1.5 **RETINAL IMPLANTS**

In retinal diseases such as retinitis pigmentosa and AMD where the photoreceptors are damaged, the inner bipolar and ganglion layers are relatively intact and still functioning. Consequently, in order to restore some form of light perception and ultimately vision perception, a retinal implant would need to focus on replicating the sensation of light and darkness by artificially hyperpolarizing and depolarizing remaining photoreceptor and subsequent bipolar cells in a damaged retina.

The technique used by many retinal implantable devices is to stimulate the inner nerve cells of the retina electrically with an ordered pattern of electrical impulses using arrays of electrodes implanted into the retina. The electrical stimuli can then be derived from extracted video data from an external video camera attached to a pair of glasses. Alternatively, microphotodiodes can be used to convert the incident light energy of images on the retina, as the lens of the eye is still functional, into electrical stimuli. The array of microphotodiodes and microelectrodes are symmetrically aligned such that they effectively bypass the outer damaged photoreceptor cells, stimulating the inner nerve cells directly. Retinal prostheses can also provide conditioned electrical impulses to evoke patterns of light dots to represent Braille characters.

Retinal implants can be epiretinal where the implant is inserted on the surface of the retina with electrodes extending into the internal layers of the retina to stimulate either the bipolar or ganglion cells; subretinal where the implant is inserted inside the retina in the photoreceptor layer; or suprachoroidal where implants are implanted in the suprachoroidal space at the back of the eye between the retina and the sclera of the eye.

Other considerations for retinal implants, apart from suitable biocompatible materials, include the mechanism by which the implant is "fixed" in place as well as the technique on how to supply power to the electronic devices in the implant. Typical techniques used include inductively coupled magnetic field coils to transfer energy from an external source to an implanted receiver coil, energy harvesting from the incident light falling on the retina, or using an external infrared laser beam mounted on a pair of glasses to power the implant.

One measure of achieved visual resolution of retinal implants is the visual acuity achieved. Visual acuity refers to the contrast and resolution detail in which an image in the center of vision can be seen. Other terms used include the sharpness, clearness, or acuteness of a perceived image. Visual acuity is measured relative to normal vision, which is defined as

20/20 and refers to the ability of the human eye to distinguish between separate arc lines drawn on a chart, 20 ft (6 m) away. Each arc line is separated by 1 min (sixtieth) of a degree (equivalent to a separation of 1.75 mm). In comparison, the detail that a person with 20/40 vision can read at a distance of 20 ft, can clearly be read by a person with 20/20 vision at a distance of 40 ft.

1.6 **MICROELECTRODES**

Cell membrane potentials can be altered by injecting small bidirectional currents into the subretinal layer to induce the sensation of light, the image resolution being dependent on the population of bipolar cells that can be activated. The delivery of these currents to a small number of cells would necessitate using micro- or nanofabrication of needle-type electrodes in order to provide adequate "pixel" resolution (retinoscopy) and sensitivity to restore some form of visual perception. However, the amount of bidirectional current that can be safely delivered is dependent on the charge "capacity" and material of the electrode. Charge capacity is defined as the maximum amount of charge per unit area that can be delivered by a biphasic current pulse to an electrode without sustaining substantial electrode damage. The smaller the electrode, the lesser its charge capacity. The actual electrode charge capacity is dependent on the applied electrode potential. The charge density at the interface of the electrode and surrounding tissue, the electrolyte, is defined as the injected charge per phase of a biphasic stimulation pulse, per unit surface area of the electrode.

As the electric charge delivered is proportional to the electrode surface area, there is a trade-off between the electrode size, to activate a sufficient number of neural cells for sufficient image resolution and the amount of charge that can be delivered without incurring electrode or surrounding tissue damage. Increasing the electrode potential can lead to the electrolysis of water in which the reduction of water, for a negative going current pulse, produces hydrogen gas, whereas a positive going current pulse results in the oxidation of water, producing oxygen gas. The production of oxygen and hydrogen gases, also known as "bubbling," is irreversible. A range of electrode potentials known as the "water window" defines the limits between which no gases are produced. The reversible charge injection can be defined as the maximum charge density applied without the electrode exceeding the water window during pulsing (Roblee and Rose, 1990). This therefore defines the charge injection capacity limit of an electrode. Other irreversible reactions include metal corrosion where an electrode is driven to a positive potential which

Table 1.1 Electrode Potential and Electrochemical Charge Limits for TiN and IrO$_x$

	Electrode Potential Limits	Electrochemical Charge Limits (mC/cm^2)
TiN	−0.75 to −1.25 V	0.6−0.9
IrO$_x$	−0.6 V to 08 V	1−3

causes the metal to oxidize, resulting in the production of toxic reactants. The requirement then is to make electrode arrays smaller but still retain a high charge density.

The microelectrode arrays for retinal implants most commonly use a metal oxide coating of either titanium nitride (TiN) or iridium oxide (IrO$_x$) resulting in a higher electrode charge capacity density compared to other noble metal electrodes such as platinum and iridium. The coatings act as a dielectric, forming a double-layer capacitor interface between the electrode and the surrounding tissue that forms the electrolyte. A negative going cathodic pulse on the electrode will induce a positive ionic current in the electrolyte toward the electrode eliciting a neural cell membrane depolarization. The subsequent positive going anodic pulse will reverse the ionic current flow, resulting in a net zero injection of charge. Table 1.1 provides the electrochemical limits for TiN and IrO$_x$.

1.7 MICROPHOTODIODES

Microphotodiodes consist of an array of individual photodiodes and stimulation electrodes such that they effectively transduce incident light energy into an electric stimulus, thus replicating the function of photoreceptors. Each microphotodiode has an electrode array configuration such that a pixelated image effectively will be mapped to stimulate the appropriate bipolar cells approaching that of the natural retinotopy of the received retinal image. The larger the array of photodiodes and electrodes, the greater the pixel (spatial) resolution and the subsequent retinotopy of the retinal implant. The photodiodes generate currents which are typically converted into a voltage. The resultant monophasic voltage pulses are applied to capacitively coupled electrodes in order to provide a biphasic charge-balanced stimulus pulse, minimizing the possible risk of cell damage which would normally result from a nonzero polarization of the electrode.

1.8 ARGUS II RETINAL PROSTHESIS (SECOND SIGHT MEDICAL PRODUCTS)

The Argus II Retinal Prosthesis from Second Sight Medical Products has been designed to help patients with loss of sight mainly due to degenerative retinal diseases such as retinitis pigmentosa where there is a degeneration of the retinal outer layer of light-sensitive photoreceptors such that the patient has bare light or no light perception.

■ **FIGURE 1.7** The Argus II Retinal Prosthesis System from Second Sight. *(Copyright © 2013 Second Sight Medical Products, Inc. Reprinted with permission.)*

The Argus II System consists of a pair of glasses which incorporates an internal video camera which captures a scene (Figure 1.7). The video data for the scene is sent by a cable to the video processing unit which is an external device normally worn on a belt and contains the battery power supply. The processed video data is then sent back to the glasses where it is transmitted wirelessly to a receiver implanted on the outer surface of the eyeball. The receiver then sends electrical pulses via a thin flat wire, through a small incision in the eye, to an electrode array which is implanted in the eye (Figure 1.8).

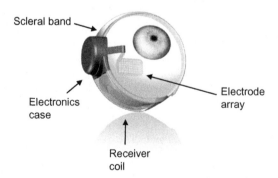

■ **FIGURE 1.8** The Argus II Retinal Prosthesis from Second Sight Medical Products. *(Copyright © 2013 Second Sight Medical Products, Inc. Adapted and printed with permission.)*

The electrode array consists of 60 electrodes on a substrate measuring 1 mm × 1 mm which is placed at the back of the eye over the retina and secured in place by a surgical tack centered on the macula region (Figure 1.9).

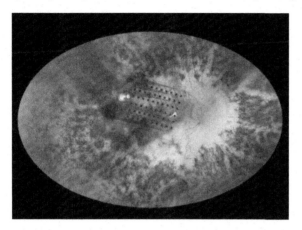

■ **FIGURE 1.9** Electrode array secured in place over the retina. *(Copyright © 2013 Second Sight Medical Products, Inc. Reprinted with permission.)*

The electrodes make contact with the inner nerve retinal ganglion layer in the retina thus bypassing the outer damaged photoreceptors. With retinitis pigmentosa, the inner nerve retinal ganglion cells are still functional and still retain a viable connection to the optic nerve such that electrical pulses applied to the electrode array will stimulate nerve fibers in the optic nerve.

Figure 1.10 shows the components of the retinal prosthesis. The scleral band wraps around the eye and goes underneath the eye muscles

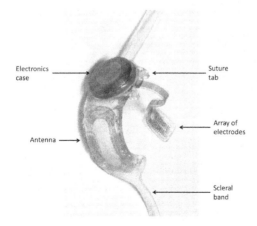

■ **FIGURE 1.10** Components of the Argus II which sit in the eye socket. *(Copyright © 2013 Second Sight Medical Products, Inc. Reprinted with permission.)*

and is held in place by sutures on the band. Embedded in the strap is the coil which receives external data and supplies power to the receiver. The whole assembly sits within the eye socket hidden from view (Figure 1.11).

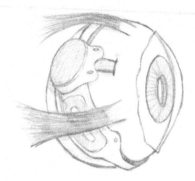

■ **FIGURE 1.11** Assembled Argus II which sits in the eye socket. *(Copyright © 2013 Second Sight Medical Products, Inc. Argus II Retinal. Reprinted with permission.)*

The Argus II has approved regulatory CE mark from Europe and FDA approval. Patients using the Argus II have reported being able to make out shapes, movement, distinguish between light and dark, and read large letters. The next generation of Second Sight implants, Argus III and IV, will incorporate a 256-grid electrode array.

1.9 ARTIFICIAL SILICON RETINA IMPLANT, OPTOBIONICS

The Artificial Silicon Retina (ASR) from Optobionics is a self-contained, self-powered retinal implant designed to convert light energy from images into electrical impulses, stimulating the nerve cells within the retina. The ASR is a microchip with approximately 5000 independently functioning microelectrode-tipped microphotodiodes. Incident light energy is also used to power the microchip. Consequently, there are no external power supplies, leads, or external camera to capture the images. The ASR measures 2 mm in diameter with a thickness of 25 µm and is implanted under the retina in the subretinal space (Figure 1.12).

Figure 1.13 shows the fabricated microphotodiode pixels, each measuring 20 µm × 20 µm with a 9 µm × 9 µm iridium oxide electrode deposited and electrically bonded to each pixel. The separation between pixels is 5 µm × 5 µm. With a light illumination of 800 foot-candles, the pixel current is between 8 and 12 nA.

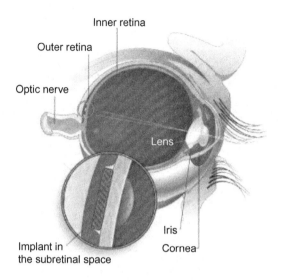

■ **FIGURE 1.12** *ASR in situ. (Copyright © Optobionics. Reprinted with permission.)*

■ **FIGURE 1.13** *ASR microphotodiode pixel (Chow et al., 2004). (Reprinted with permission.)*

1.10 **ALPHA-IMS IMPLANT BY RETINA IMPLANT AG**

The Alpha-IMS is a subretinal prosthesis that converts light energy from retinal images into electrical signals to effectively bypass damaged photoreceptor cells and stimulate still active bipolar neural cells. The implant is placed beneath the fovea in the outer retina and is powered from an external power source providing inductively coupled energy (and control signals) to a receiver coil implanted under the skin behind the ear. The external

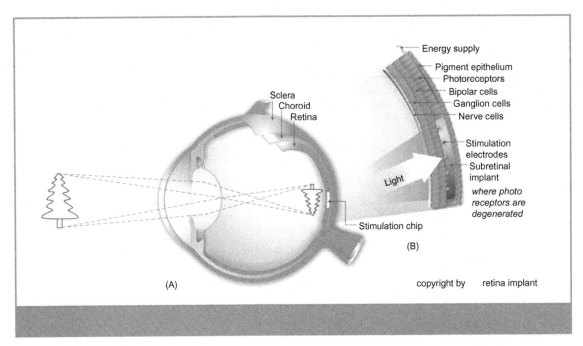

■ **FIGURE 1.14** (A) Alpha-IMS subretinal prostheses. (B) Implant inserted in the subretinal space toward the outer retina. *(Copyright © Retina Implant AG. Reprinted with permission.)*

transmitter coil is held in place by magnets inserted into the skull. A silicone cable from the receiver coil runs subdermally to the eye socket where it connects to a thin polyimide foil cable which enters the subretinal space through a small incision in the sclera and choroid at the back of the eye (Figure 1.14). A reference electrode is placed subdermally in the orbit rim of the eye.

Figure 1.15A shows the implant which consists of a microchip placed on a flexible polyimide printed circuit board. The microchip contains an array of 1500 photodiodes and 1500 electrodes arranged as individual elements as shown in Figure 1.15B. Each element contains a rectangular $15\,\mu m \times 30\,\mu m$ microphotodiode, a $50\,\mu m \times 50\,\mu m$ TiN electrode array, and a differential amplifier, each element measures $72\,\mu m \times 72\,\mu m$. The electrodes are $70\,\mu m$ long with a diameter of $70\,\mu m$ and are spaced $70\,\mu m$ apart. The implant also contains an array of 4×4 test electrodes for direct stimulation of bipolar cells. The test electrodes are light independent and provide a means to assess electrode interface characteristics and current pulse stimulation patterns. The microchip measures $3\,mm \times 3\,mm$ with a thickness of $70\,\mu m$ and is fabricated using $0.8\,\mu m$ CMOS technology.

The biphasic stimulation current delivered by the electrodes is dependent on the amount of light striking the microphotodiodes and typically delivers

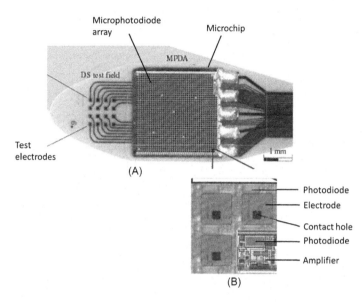

■ **FIGURE 1.15** Alpha-IMS prostheses: (A) microchip and direct stimulation electrode array and (B) microphotodiode element. *(Copyright © Retina Implant AG. Reprinted with permission.)*

between 20 and 60 nC of charge per stimulation pulse. The light-to-voltage conversion period is between 0.5 and 6 ms and the pixel image refresh rate is typically 5–7 images per second. The microchip array covers an estimated visual angle of 11° × 11° with 1° representing a distance of 288 μm on the retina. The Alpha-IMS has been awarded the European CE mark.

1.11 **BIONIC VISION AUSTRALIA**

There are three implantable devices from Bionic Vision Australia (BVA), the Early prototype (24 electrodes), the Wide-View device (98 electrodes), and the High-Acuity device (1024 electrodes). The Early prototype and Wide-View implants are inserted into the suprachoroidal space, whereas the High-Acuity incorporates an epiretinal implant inserted onto the surface of the retina. Image data from an external camera located on the frame of the user's glasses is processed and sent to a microchip-electrode array.

Figure 1.16 shows the Wide-View BVA, 98 electrode system. Images captured by the camera are sent to an external processing unit where the data is processed and sent to the implanted system via a wired connection at the back of the ear which connects to the microchip-electrode array.

With the High-Acuity BVA system, Figure 1.17, the image data captured by the external camera is wirelessly transmitted to the retinal implant where the processor analyzes the data and delivers the appropriate electrode

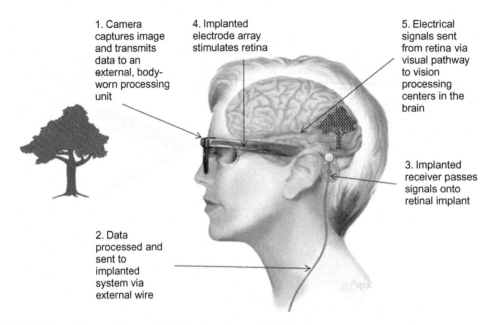

1. Camera captures image and transmits data to an external, body-worn processing unit

4. Implanted electrode array stimulates retina

5. Electrical signals sent from retina via visual pathway to vision processing centers in the brain

3. Implanted receiver passes signals onto retinal implant

2. Data processed and sent to implanted system via external wire

■ **FIGURE 1.16** Wide-View BVA system. *(Courtesy of Bionic Vision Australia, copyright Beth Croce.)*

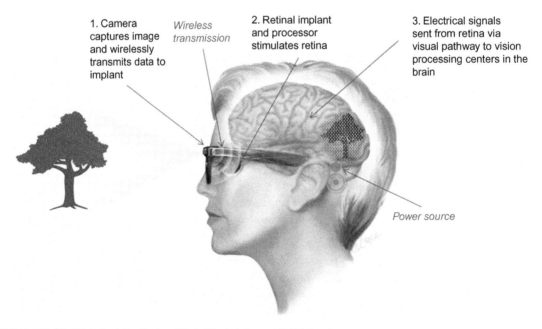

1. Camera captures image and wirelessly transmits data to implant

Wireless transmission

2. Retinal implant and processor stimulates retina

3. Electrical signals sent from retina via visual pathway to vision processing centers in the brain

Power source

■ **FIGURE 1.17** BVA High-Acuity device. *(Courtesy of Bionic Vision Australia, copyright Beth Croce.)*

stimulation pattern. The 1024 microchip-electrode array consists of man-made polycrystalline diamond electrodes. The same material is used to coat the epiretinal implant which is inserted onto the surface of the retina.

1.12 BOSTON RETINAL IMPLANT PROJECT: BIONIC EYE TECHNOLOGIES, INC. AND VISUS TECHNOLOGIES, INC.

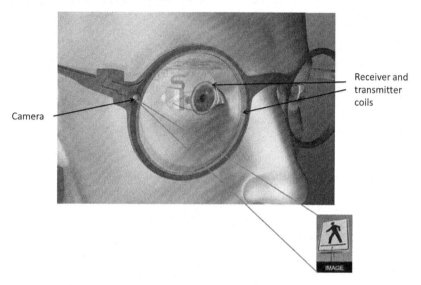

Camera

Receiver and transmitter coils

■ **FIGURE 1.18** Boston Retinal Implant Project. *(Adapted from Kelly et al. 2013. Reprinted with permission.)*

Figure 1.18 shows an overview of the system components in the Boston Retinal Implant Project. The data from an image, captured by the camera located in the special glasses worn by the user, is wirelessly transmitted to an internal receiver attached to the outside of the eye and converted by the signal processor into current pulses which are delivered by a fine wire to the electrode array implanted in the subretinal space.

The implant consists of hermetically sealed titanium enclosure measuring $11 \text{ mm} \times 11 \text{ mm} \times 2 \text{ mm}$ which sits on the outside of the eye and is sutured to the sclera. Power and data is wirelessly transferred using a near-field inductive coupling at a frequency of 6.75 MHz using Frequency Shift Keying achieving a data transmission rate of 565 kbps. Power for the implant is derived from a pair of transmitter and receiver coils. The receiver coil is located on the front of the eye around the limbus, the border area of the cornea and sclera (Figure 1.19), whereas the transmitter coil is integrated into the frame of a pair of "Smart glasses" worn by the user.

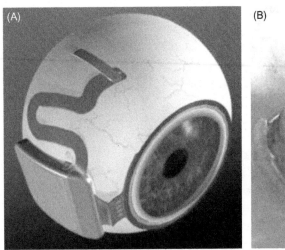
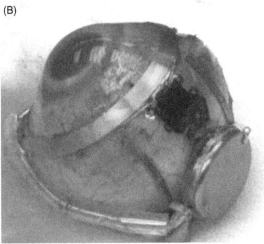

■ **FIGURE 1.19** Site of receiver enclosure, receiver coil, and wire link to the electrode array in (A) model concept and (B) a human eye (Kelly et al., 2013). *(Reproduced with permission.)*

The serpentine connector extends from the implant enclosure to the electrode array which is inserted into the subretinal space. The electrode array consists of 256 microfabricated Sputtered Iridium Oxide Film needle-type electrodes of diameter 400 μm on a thin-film polyimide substrate base. The custom designed ASIC is capable of delivering a maximum stimulating current of 127 μA in steps of 1 μA with a maximum pulse width of 4500 μs in steps of 17.7 μs. To allow for the variations in electrode sizes, charge limits of 50, 100, and 200 nC can be set. The stimulating electrodes can also be configured as current sources or sinks to allow for current steering and there is a provision to measure and minimize residual electrode voltage after delivery of a stimulus pulse. Over time, a residual DC voltage can damage the electrodes and surrounding tissue. The whole ASIC is fabricated using a 0.18 μm CMOS process.

BIBLIOGRAPHY

Chow, A.Y., Chow, V.Y., Packo, K.H., Pollack, J.S., Peyman, G.A., Suchard, R., 2004. The artificial silicon retina microchip for the treatment of vision loss from retinitis pigmentosa. Arch. Opthalmol. 122 (4), 460–469.

Kelly, S.K., et al., 2013. Developments on the Boston 256-channel retinal implant. Proc. IEEE International Conference on Multimedia and Expo Workshops, pp. 1–6.

Roblee, L.S., Rose, T.L., 1990. Electrochemical guidelines for selection of protocols and electrode materials for neural stimulation. Neural Prostheses: Fundamental Studies. Prentice-Hall, Englewood Cliffs, NJ, pp. 25–66.

Stett, A., et al., 2000. Electrical multisite stimulation of the isolated chicken retina. Vision Res. 40 (13), 1785–1795.

Chapter 2

Smart Contact Lens

2.1 INTRODUCTION

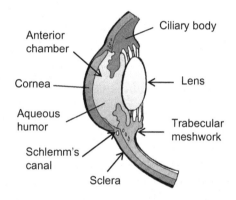

■ **FIGURE 2.1** Anterior structure of the eye.

Glaucoma is a progressive disease resulting from an increase in the internal pressure of the eyeball which can damage the optic nerve at the back of the eye. In a normal eye, there is a dynamic equilibrium of the internal aqueous humor fluid which maintains the shape of the eye. Aqueous fluid is secreted by cells in the ciliary body (Figure 2.1), which is located behind the iris of the eye in the posterior chamber. The aqueous humor fluid passes over the lens and through the pupil into the anterior chamber in front of the iris. The aqueous humor then leaves the eye by a network of microscopic drainage channels consisting of the trabecular meshwork and Schlemm's canal into the episcleral veins to be absorbed into the bloodstream. A blockage of the drainage channels, an increase in the production of aqueous humor fluid, or an increase in the episcleral venous pressure can all lead to an increase in the level of the intraocular pressure (IOP) which will damage the axons in the optic nerve. The aqueous humor is not to be confused with "tears" which spread over the outer

Implantable Electronic Medical Devices. DOI: http://dx.doi.org/10.1016/B978-0-12-416556-4.00002-4

surface of the cornea or the vitreous humor fluid in the posterior chamber of the eye behind the lens.

The effects of glaucoma are irreversible, eventually leading to loss of sight. However, if detected early, the onset of the disease can be managed with medical treatment or laser surgery. Measuring the IOP of the eye can help in detecting the early stages of the disease.

2.2 **MEASUREMENT OF IOP**

One way to measure the IOP of the eye is to measure the amount of counterpressure (push back) that a small area of the cornea exerts when flattened by an external pressure. This can be done using tonometer instruments, using direct cornea contact or noncontact cornea applanation (flattening) techniques. Contact applanation tonometers such as the Goldmann tonometer measures the force exerted by a probe tip pushed against a fixed area of the cornea, whereas a typical noncontact tonometer measures the force exerted by small puffs of air to flatten a fixed area of the cornea in order to determine the IOP. The Goldmann tonometer is widely recognized as the "gold" standard test when measuring IOP.

Based on the initial version of Imbert–Fick law, a force applied against a sphere equals the pressure in the sphere multiplied by the area flattened (applanated) by the external force such that:

$$W = Pt \times A \tag{2.1}$$

where

W—applied force (g)
Pt—pressure in the sphere (mmHg)
A—area flattened (applanated) by the external force (mm^2).

The Imbert–Fick law assumes that for a perfect sphere, the membrane is infinitely thin, perfectly flexible, and dry. However, the rigidity of the cornea will offer some initial resistance which the applied force will have to overcome. The cornea is also wet, having a thin "tear" film creating a surface tension which creates a capillary attraction, pulling the tonometer probe tip in toward the cornea, as shown in Figure 2.2.

The modified Imbert–Fick law then becomes:

$$W + S = (Pt \times A) + B \tag{2.2}$$

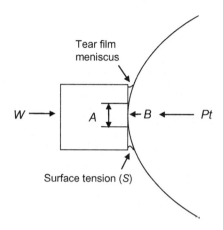

■ FIGURE 2.2 Forces applied to applanated area of cornea.

where

S—surface tension of tear film meniscus

B—force required to bend cornea (overcome cornea rigidity).

Goldmann (Moses, 1958) found that when flattening the cornea to within a diameter range of 2.5–4 mm, the surface tension of tears and corneal rigidity forces cancel out. At a diameter of 3.06 mm, the ratio of the IOP to the applied gram-force was found to be 10:1 (Duke-Elder, 1968; Davson, 1962; Moses, 1958), i.e., 1 mmHg of IOP equals one-tenth of the applied flattening force in grams, which provides a convenient conversion factor. Therefore, for an IOP of 15 mmHg, a force of 1.5 g is required to flatten the cornea to a diameter of 3.06 mm. The diameter of 3.06 mm corresponds to a flattening area of 7.36 mm². Normal eye pressure of the anterior chamber is defined to be in the range of 10–21 mmHg (Pinsky and Dayte, 1991).

In the case of contact tonometers, anesthetic eye drops are instilled into the eye, and in the case of the Goldmann tonometer, a fluorescein dye is instilled into the eye which causes the tear meniscus to fluoresce bright yellowish under blue cobalt light which when applied to two separate prisms, produces two separate semicircle images known as mires. The applied force of the probe tip is increased until the semicircles are moved apart until they are just touching. This represents a flattened cornea diameter of 3.06 mm. In order to build up a complete IOP profile, a number of tests would be required in order to detect any IOP elevations which may occur during the course of daytime and nighttime.

Noncontact tonometers include the air puff tonometer in which a puff of air with a known contact area is directed at the surface of the cornea with a force which increases linearly with time, progressively flattening the cornea until it acts like a mirror reflecting a beam of light onto a light sensor, turning off the applied puff of air. Knowing the applied force of the puff of air and the cornea contact area, the IOP can be calculated.

The IOP can be measured by using "smart" contact lenses which detect corneal deformations due to changes in the IOP. These deformations can be detected using microfabricated strain gauges or resonance frequency selective electrodes. Other techniques include placing sensors inside the anterior chamber of the eye.

2.3 **TRIGGERFISH FROM SENSIMED**

The Triggerfish IOP sensor from Sensimed is a "smart" contact lens designed to monitor IOP changes in the eye over a 24 hour period by detecting changes in the curvature of the cornea. A microfabricated strain gauge sensor, antenna, and dedicated telemetry circuits (ASIC) are embedded into the soft contact lens, arranged in such a way that they do not interfere with normal vision (Figure 2.3). A receiving antenna is taped around the eye and connected to a portable data recorder. After 24 hours the contact lens is disposed of and the data profiling of the IOP is analyzed.

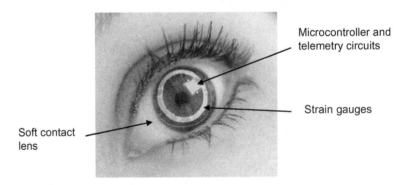

■ **FIGURE 2.3** Triggerfish smart contact lens. *(Copyright © 2013 Sensimed. Reprinted with permission.)*

As the IOP increases, there is an outward spherical distribution of forces on the cornea, as shown in Figure 2.4. The contact lens will stay in contact with the cornea due to the capillary forces of the pre- and post-tear films and will stretch following deformations of the cornea. The soft lens will therefore expand with an increase in the corneal radius of curvature.

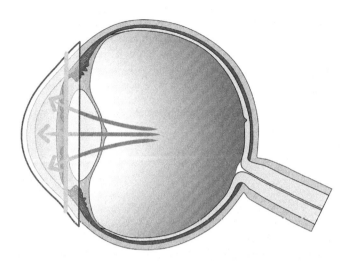

■ **FIGURE 2.4** Increase in IOP on contact lens. The vertical line indicates the position of the circumferential strain gauges. *(Copyright © 2013 Sensimed. Reprinted with permission.)*

The subsequent circumferential change of the contact lens is detected by the circular strain gauges with a nominal diameter of 11.5 mm corresponding to that of the corneoscleral junction, which is where the most significant circumferential strains can be detected. Figure 2.5 shows an exaggerated change in the corneal radius as the cornea is pushed outward with an increase in the IOP. It is assumed that the maximum corneal deformation will occur at the corneoscleral junction, also known as the limbus transition zone, which is the boundary between the cornea and the sclera regions of the eye.

The soft contact lens consists of a polyimide microflex substrate onto which platinum–titanium metallic strain gauges patterns are

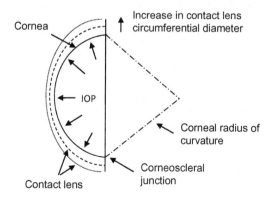

■ **FIGURE 2.5** Change in circumferential diameter of contact lens with a change in IOP.

microfabricated and then sandwiched between two layers of polyimide. The polyimide layers are then embedded into the silicone of a soft contact lens. The ASIC is a custom-built IC which provides the control and signal conditioning circuits together with the radio telemetry circuits.

The detected strain is given by (Leonardi et al., 2003):

$$\varepsilon = \frac{\Delta R}{GFR0} \tag{2.3}$$

where

ε—strain of the active gauges
GF—strain gauge factor
R0—nominal strain gauge resistance
ΔR—change in strain gauge resistance.

Using a Wheatstone bridge configuration with two active arms (circumferential strain gauges) and two passive arms (fixed strain gauges), the output voltage dependent on the strain is given by:

$$Vm = \frac{\varepsilon GFVs}{2} \tag{2.4}$$

where

Vm—measured output voltage
Vs—applied DC voltage
GF—strain gauge factor

The resultant measured strain ε does not vary linearly with changes in the curvature Δr, of the cornea. However, the change in the central corneal curvature radius is significantly less than the nominal corneal curvature radius such that the strain versus corneal radius curvature can be considered linear. Typical values for the change in corneal curvature vary between 1 and 200 μm for a corneal curvature radius of 8 mm (Hjortdal and Jensen, 1995; Lam and Douthwaite, 1997).

The strain can therefore be given as:

$$\varepsilon = a\Delta r \tag{2.5}$$

where

Δr—change in the curvature of the contact lens
a—proportional factor.

The measured output signal from the strain gauge can then be rewritten as:

$$Vm = \frac{a\Delta r \text{GFV}s}{2} \tag{2.6}$$

The IOP is a function of the corneal curvature such that the change in pressure can be defined as:

$$\Delta p = b\Delta r \tag{2.7}$$

Therefore, substituting for Δr from Eq. (2.7) into Eq. (2.6), the change in IOP is given by:

$$\Delta p = \frac{2bVm}{a\text{GFV}s} \tag{2.8}$$

Substituting for ε from Eq. (2.3) into Eq. (2.4) and rearranging:

$$\Delta R = \frac{2VmR0}{Vs} \tag{2.9}$$

Hence, ΔR can be determined from the measurement of Vm from the Wheatstone bridge with a known applied Vs.

Assuming that a 1 mmHg increase in IOP causes a 3 μm change in the corneal radius of curvature of nominal radius 7.8 mm (Hjortdal and Jensen, 1995; Lam and Douthwaite, 1997), the IOP can be determined such that "b" in Eq. (2.7) can be calculated as:

$$b = \frac{\Delta p}{\Delta r} = \frac{1 \text{ mm}}{3 \text{ μm}} = \frac{1}{3}\text{mm}$$

BIBLIOGRAPHY

Davson, H., 1962. The Eye. Vegetative Physiology and Biochemistry, vol. 1, third ed. Academic Press, San Diego, CA.

Duke-Elder, S., 1968. The physiology of the eye and of vision. System of Opthalmology, vol. 4. C.V. Mosby, St. Louis, MO, p. 157.

Hjortdal, J., Jensen, P.K., 1995. In vitro measurement of corneal strain, thickness and curvature using digital image processing. Acta Ophthalmol. Scan. 73, 5−11.

Lam, A.K., Douthwaite, W.A., 1997. The effect of an artificially elevated intraocular pressure on the central corneal curvature. Ophthal. Physiol. Opt. 17, 18−24.

Leonardi, et al., 2003. A soft contact lens with a MEMS strain gage embedded for intraocular pressure monitoring, 12th International Conference on Solid State Sensors, Actuators and Microsystems, Boston, MA, pp. 8−12.

Moses, R.A., 1958. The Goldmann applanation tonometer. Am. J. Opthalmol. 46−865.

Pinsky, P.M., Dayte, D.V., 1991. A microstructurally-based finite element model of the incised human cornea. J. Biomech. 24 (10), 907−909, 911−922.

Chapter

Phrenic Nerve Stimulation

3.1 INTRODUCTION

Phrenic nerve stimulation, also known as phrenic nerve pacing, has been used to restore some form of breathing function in patients with respiratory paralysis resulting from spinal cord injuries or from neurological disorders such as congenital central hypoventilation syndrome, central sleep apnea, and diaphragm paralysis. Phrenic nerve pacing has also been shown to be beneficial for patients with motor neuron diseases such as amyotrophic lateral sclerosis, also known as Lou Gehrig's disease, to assist with breathing, sleeping, and delaying the time for dependence on a mechanical ventilator.

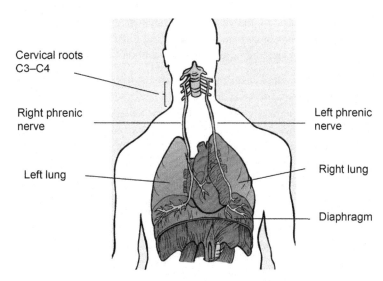

■ **FIGURE 3.1** Innervation of the diaphragm.

The phrenic nerve originates in the spinal cervical roots, C3, C4, and C5, and innervates the diaphragm which is a sheet of skeletal muscle attached to the lower rib cage, forming a dome-shaped base to the thoracic cavity containing the heart, lungs, and rib cage (Figure 3.1). When the diaphragm contracts, it

Implantable Electronic Medical Devices. DOI: http://dx.doi.org/10.1016/B978-0-12-416556-4.00003-6

begins to flatten and move downward, increasing the volume of space in the thoracic cavity above the diaphragm. This results in a decrease in pressure inside the lungs, causing a negative differential pressure with respect to the atmospheric pressure, thus drawing air into the lungs, which is known as inspiration. When the diaphragm relaxes, it moves back up into the thoracic cavity, increasing the pressure and forcing air out of the lungs (expiration).

Trauma or neurological damage to the cervical roots will result in impaired or loss of breathing to an extent that the standard treatment is the use of a mechanical ventilator which forces air into the lungs through an endotracheal tube or tracheostomy tube under positive pressure. However, there is a risk of infection after a patient is intubated as bacteria and other microorganisms can enter the endotracheal tube and infect the lungs. Ventilator-associated pneumonia is defined if pneumonia is detected after 48 hours after intubation.

Mechanical ventilators will also affect the level of independence and mobility a patient has and can also impair speech. The inability to cough means that lung infections can lead to a build-up of fluid in the lungs and there can also be muscle atrophy of the diaphragm due to nonusage. However, if the phrenic nerve is intact and the diaphragm muscles are still functioning, electrical stimulation of the phrenic nerve or electric stimulation of the muscle motor points of the muscles of the diaphragm can produce a contraction of the diaphragm muscles causing air to be taken into the lungs under negative pressure, thus affecting a more natural breathing action.

Diaphragm pacing relies upon the electrical stimulation of muscle motor points of the diaphragm using intramuscular electrodes, whereas phrenic nerve stimulation utilizes electrodes placed in contact with the nerve. In either case, the efficiency of the implantable system can be determined by measuring the inspiration time when the diaphragm contracts, the tidal volume (depth of breath), and the effective respiratory rate that can be achieved. The respiratory rate is defined as the number of breaths taken over a period of 60 s where one breath represents one complete inhalation–exhalation cycle. The average resting respiratory rate decreases with age but is considered to be between 12 and 18 breaths per minute (bpm) in adults with a tidal volume of around 500 mL (Blows, 2001).

The phrenic nerve can be accessed using different cervical and thoracic approaches to expose the phrenic nerve in the neck region and even through the second intercostal space (between the second and third rib) of the rib cage. Minimally invasive surgical techniques are normally used for implantation of phrenic nerve electrodes in the thoracic cavity and intramuscular electrodes in the abdomen for diaphragm pacing. After implantation, a conditioning electrical stimulation of the diaphragm is usually performed in order to restore muscle endurance and strength after atrophy and to condition the musculature of the diaphragm for pacing.

3.2 **ATROTECH ATROSTIM PHRENIC NERVE STIMULATOR**

The Atrostim Phrenic Nerve Stimulator (PNS) is a diaphragm pacing system which consists of two implanted stimulators, each connected to a multipolar electrode configuration placed in contact with the left and

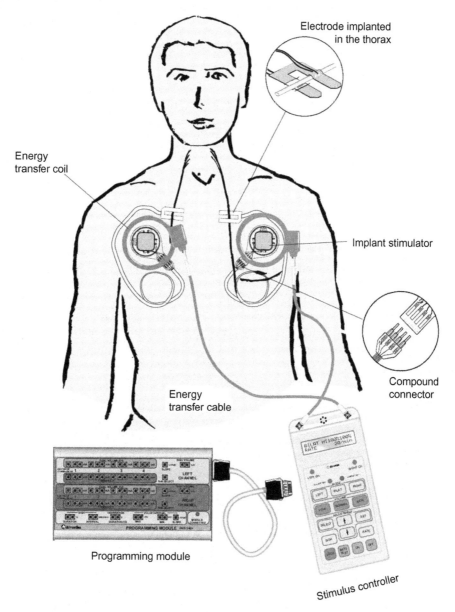

■ **FIGURE 3.2** Atrostim PNS system. *(Copyright © Atrotech. Reprinted with permission.)*

right phrenic nerves, respectively. The stimulators are implanted subcutaneously in the upper chest region and incorporate an inbuilt pickup coil for power and communication. An external controller inductively transmits energy and data to the implants via transmitter coils attached to the skin over the sites of the implants (Figure 3.2).

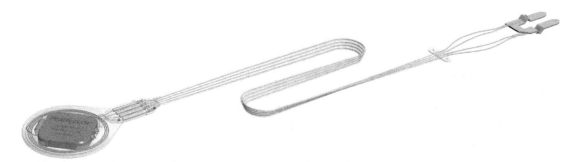

■ FIGURE 3.3 Atrostim receiver and four-contact electrode. *(Copyright © Atrotech. Reprinted with permission.)*

The multipole electrode (Figure 3.3) consists of four platinum contact electrodes which effectively divide the stimulation site into four quadrants. Each individual quadrant stimulus will result in partial muscle contractions of the diaphragm. However, stimulating all four quadrants sequentially will affect a fused muscle contraction at a quarter of the stimulus frequency (4.5–5.5 Hz) compared to the stimulus frequency (18–22 Hz) using a monopolar electrode. The sequential stimulation will also allow for a greater recovery period of the muscle motor units compared to monopolar stimulation such that the patient's diaphragm tolerates continuous stimulation without fatigue and the time needed to wean a patient off a mechanical ventilator (diaphragm conditioning) is reduced.

In order to cater for different metabolic patient needs, the Atrostim PNS provides three preset stimulation settings for when the patient is in the supine position, sitting down, or has an immediate need for an increased metabolic rate. The Atrostim PNS also has manual and automatic functions for diaphragm muscle training and to assist coughing.

3.3 **AVERY BIOMEDICAL DEVICES BREATHING PACEMAKER SYSTEM**

The Breathing Pacemaker System from Avery Biomedical Devices can help restore some form of breathing function for adult and pediatric patients who have lost neurological control of respiration. The Avery system stimulates the phrenic nerves, resulting in contraction of the hemidiaphragms, causing air to be taken into the lungs under negative pressure for the inspiratory phase of the respiratory cycle. The expiratory phase and subsequent relaxation of the hemidiaphragms occurs when the stimulus is turned off. Figure 3.4 shows a standard stimulus pattern for one breath cycle with a fixed 1.3 s inspiratory period (IP) consisting of 26 μs × 150 μs-wide stimulus pulses with a pulse interval of 50 ms. The expiratory period (EP) varies depending on the respiratory rate; for example, for a respiratory rate of 12 bpm, the breath cycle is 5 s, giving an EP of 3.7 s. The IP can be preadjusted in the factory if requested by the clinician.

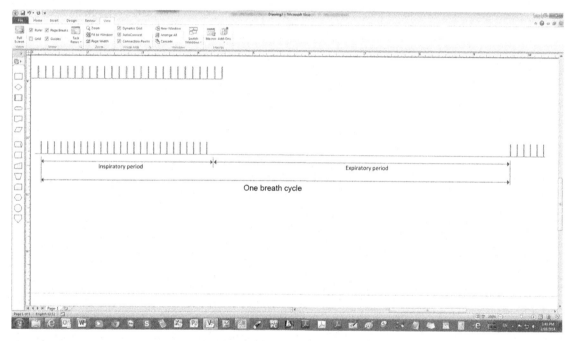

■ **FIGURE 3.4** Stimulus pattern for one breath cycle. *(Copyright © Avery Biomedical Devices. Reprinted with permission.)*

In cases when there is an abrupt contraction of the diaphragm, the airway may collapse, causing a form of obstructive apnea which requires the stimulus parameters to be changed. Another way is to gradually increase the amplitude of the stimulus pulses in order to condition the diaphragm contractions to more closely resemble a normal breath. The stimulus parameters for the current sources in the receiver are summarized in Table 3.1.

Table 3.1 Stimulus Parameters for the Avery Biomedical Devices Breathing Pacemaker System

Parameter	Minimum	Factory Setting	Maximum
Pulse width (μs)	70	150	1000
Pulse interval (ms)	40	50	130
Inspiratory time (s)	1.2	1.3	1.45
Respiratory rate (bpm)	6	12	30
Amplitude range	0 V	Dial setting 5	$8.0 \approx 10.5$ V[a]
Amplitude range (CE)	0 V	Dial setting 5	$8.0 \approx 13.5$ V[b]
Amplitude slope (V)	0	0	3.5
RF carrier frequency (MHz)	2.00	2.05	2.1

[a]FDA approved US model.
[b]CE approved EU model.

The amplitude voltages in Table 3.1 relate to the voltage drop of the stimulus current across a 1-kΩ resistor, equivalent to the internal resistance of the human body.

The Breathing Pacemaker System from Avery Biomedical Devices consists of two RF powered implantable receivers, two implantable platinum electrodes, and an external transmitter. The electrode is attached around the phrenic nerve and the receiver is surgically implanted in a subcutaneous pocket in the chest, preferably over the rib cage over which an external antenna is placed and fixed to the skin. The external transmitter is worn in a pouch and strapped to the body, wheelchair, etc. (Figure 3.5).

The transmitter (Figure 3.6A) allows the user to dial up respiratory rates from 6 to 30 bpm and can deliver stimulus currents up to 10.5 mA for the US model and 13.5 mA for the EU model. The transmitter incorporates independent circuits and batteries for the left and right sides such that if one of the batteries dies or one of the circuits fails, the other side continues to work. This redundancy guarantees pacing at all times, preventing injury or even death.

The transmitter delivers the electrical stimulus pattern to the receiver (Figure 3.6B) via the inductively coupled RF link from the antenna to the receiver coils. As the receiver does not contain batteries, it converts the RF signal into power and electrical stimulus pulses. Subsequently, if no RF signal is received, the internal components are not powered and hence no stimulus is generated.

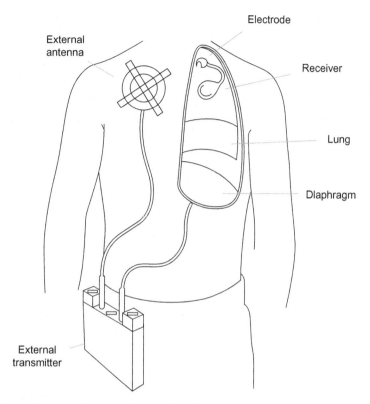

■ FIGURE 3.5 The Avery Biomedical Devices breathing pacemaker system. *(Copyright © Avery Biomedical Devices. Reprinted with permission.)*

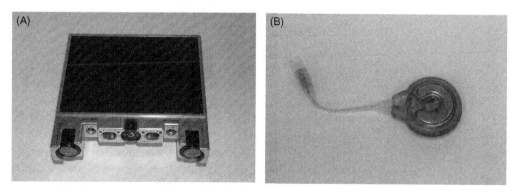

■ FIGURE 3.6 The breathing pacemaker system: (A) transmitter and (B) RF receiver. *(Copyright © Avery Biomedical devices. Reprinted with permission.)*

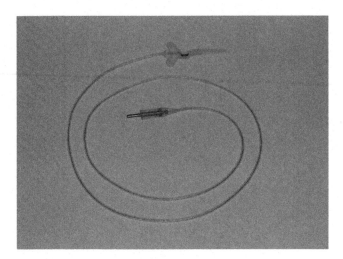

■ **FIGURE 3.7** Unipolar phrenic nerve electrode. *(Copyright © Avery Biomedical devices. Reprinted with permission.)*

The implanted electrode is made of a highly flexible stainless steel wire, insulated by silastic tubing with a platinum nerve contact on one end and a stainless steel male connector that mates with the receiver at the other end (Figure 3.7). The silastic body of the electrode connector has a small rib to make a tight closing with the receiver connector boot and to prevent intrusion of fluids that could potentially cause corrosion. The lead wire of the electrode and receiver has a diameter of about 0.3 mm and is made of 300 strands of stainless steel.

3.4 **SYNAPSE BIOMEDICAL INC. NeuRx DIAPHRAGM PACING SYSTEM**

The NeuRx Diaphragm Pacing System (DPS) provides electrical stimulation of the diaphragm muscles to cause diaphragm contractions to affect air to be taken into the lungs under negative pressure. The DPS system consists of four implantable Permaloc intramuscular electrodes inserted into the muscles of the diaphragm and a reference electrode which is placed under the skin. The wires from the electrodes exit the skin via a percutaneous electrode connector which plugs into an external four-channel stimulator (pulse generator). As well as providing stimulation of muscle motor points, the NeuRx External Pulse Generator (Figure 3.8) reads and displays impedance relative to circuit resistance. An "X" signifies a break in the specific circuit. The patient's breath per minute is adjusted until the natural rate and synchronicity is established. The stimulation consists of capacitively coupled, biphasic current pulses. The NeuRx electrode is shown in Figure 3.9.

■ **FIGURE 3.8** NeuRx External Pulse Generator. *(Copyright © Synapse Biomedical Inc. Reprinted with permission.)*

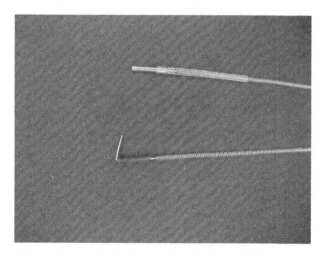

■ **FIGURE 3.9** NeuRx electrode. *(Copyright © Synapse Biomedical Inc. Reprinted with permission.)*

BIBLIOGRAPHY

Blows, W.T., 2001. The Biological Basis of Nursing: Clinical Observation. Routledge, London.

Chapter 4

Glucose Biosensors

4.1 INTRODUCTION

Glucose is a sugar derived from the digestion of carbohydrates in the body which is absorbed into the bloodstream and circulated throughout the body for cells to convert into energy. The blood glucose levels are maintained by insulin, which is a regulatory hormone produced in the pancreas by the islets of Langerhans and is required by cells for the uptake of glucose from the blood. The glucose is metabolized by cells, providing energy essential for the regulation and balance (homeostasis) of the biological systems that make up the human body.

Failure of the body to produce insulin results in diabetes mellitus. Type 1 is an autoimmune disease whereby the islet cells of the pancreas are destroyed, leading to insufficient or no insulin being produced and resulting in a high concentration level of glucose in the blood. Type 2 occurs when insulin is produced but is ineffective in "signaling" cells (insulin resistance) to uptake glucose from the blood. This form is often the result of lifestyle choices such as poor diet and lack of exercise.

High concentrations of blood glucose levels (hyperglycemia) can be fatal as cells are starved of sugars which are normally metabolized for energy. Instead, fats and proteins are metabolized producing toxic substances which in high levels can be fatal. Consequently, type 1 diabetics have to monitor their blood glucose levels and manage "self-injection" of insulin throughout the day, especially before a meal or a snack, to maintain their blood glucose level. The small drop of blood, from a pinprick on the finger, is applied to a disposable finger strip which is inserted into a blood glucose meter to give a reading of the blood glucose concentration level which is measured in either mmol/L or mg/dL:

> mmol/L refers to the number of molecules in one liter (millimoles/liter)
> mg/dL refers to the mass of molecules in one-tenth of a liter (milligrams/100 milliliters)

Implantable Electronic Medical Devices. DOI: http://dx.doi.org/10.1016/B978-0-12-416556-4.00004-8

where

$$1 \text{ mmol/L} = 0.5555 \text{ mg/dL and } 1 \text{ mg/dL} = 18.0 \text{ mmol/L}$$

Hyperglycemia has been defined by the World Health Organization (WHO) as to when the blood glucose level is greater than 7 mmol/L (126 mg/dL) when fasting (preprandial) and greater than 11.1 mmol/L (200 mg/dL) 2 h after a meal (postprandial). There are cases in which people have high blood glucose levels above the normal range but below the high limit for diabetes. This is known as impaired glucose tolerance (IGT) and is considered a precursor to type 2 diabetes. The WHO has defined people with IGT as those with a blood glucose level of more than 7.8 mmol/L but less than 11.1 mmol/L postprandial. There is no international agreed limit which clearly defines the lower limit of normal glucose levels below which defines hypoglycemia. Healthy individuals can have lower than average glucose levels without any adverse effects or symptoms of the disease, and the effects can vary with age. However, glucose levels below 2 mmol/L can result in unconsciousness.

In order to determine whether a person has diabetes, two tests are usually performed: impaired fasting glycemia (IFG), also known as a prediabetes test where the glucose level is measured after 8 h of fasting, and the IGT test, 2 h after ingesting a known concentration of glucose. The guidelines for the recommended glucose levels vary between different organizations and are summarized in Tables 4.1−4.4. Although there is a variation in the recommended blood glucose concentration levels from different

Table 4.1 World Health Organization Recommended Target Blood Glucose Levels for the Diagnosis of Diabetes (WHO, 2006)

Before Meals (Preprandial) mmol/L (mg/dL)	2 h After Meals (Postprandial) mmol/L (mg/dL)
≥7.0 (126)	≥11.1 (200)

Table 4.2 International Diabetes Federation Guidelines for Blood Glucose Levels (IDF, 2007)

	Before Meals (Preprandial) mmol/L (mg/dL)	2 h After Meals (Postprandial) mmol/L (mg/dL)
Nondiabetic	4.4−6.1 (82 to 110)	Under 7.8 (140)
Type 2	4.0−7.0 (72 to 126)	Under 8.5 (153)
Type 1	4.0−7.0 (72 to 126)	Under 9 (162)
Children with type 1	4.0−8.0 (72 to 144)	Under 10 (180)

Table 4.3 American Diabetes Association Recommended Target Blood Glucose Levels (ADA, 2013)

	Before Meals (Preprandial) mmol/L (mg/dL)	2 h After Meals (Postprandial) mmol/L (mg/dL)
Nondiabetic	3.9–7.2 (70–130)	Under 10.0 (180)[a]
Type 2	3.9–7.2 (70–130)	Under 10.0 (180)[a]
Type 1	3.9–7.2 (70–130)	Under 10.0 (180)[a]
Children with type 1 (0–6 years)	5.6 to 10 (100–180)	6.2 to 11.1 (110–200)[b]
Children with type 1 (6–12 years)	5.0–10.0 (90–180)	5.6 to 10 (100–180)[b]
Children with type 1 (13–19 years)	5.0–7.2 (90–130)	5 to 8.3 (90–150)[b]

[a]*Summary of glycemic recommendations for many nonpregnant adults with diabetes.*
[b]*Plasma blood glucose and A1C goals for type 1 diabetes by age group.*

Table 4.4 National Institute for Health and Care Excellence (NICE) Recommended Target Blood Glucose Levels (http://www.diabetes.co.uk/diabetes_care/blood-sugar-level-ranges.html)

	Before Meals (Preprandial) mmol/L (mg/dL)	2 h After Meals (Postprandial) mmol/L (mg/dL)
Nondiabetic	4.0–5.9 (70–106)	Under 7.8 (140)
Type 2	4.0–7.0 (70–126)	Under 8.5 (153)
Type 1	4.0–7.0 (70–126)	Under 9.0 (162)
Children with type 1	4.0–8.0 (70–144)	Under 10.0 (180)

organizations, the recommendations of the WHO in Table 4.1 provide an accepted general target range for the detection of diabetes.

Glucose concentration in the blood can be measured by sensors placed under the skin (subcutaneous) which measure the level of glucose concentration in the interstitial fluid, the fluid in which cells are immersed and surrounded. A more direct and faster measurement can be achieved by placing a sensor in the large-diameter blood vessel in the right atrium junction of the heart. Nanosensor technology provides an alternative technology whereby nanoparticle polymer beads containing a fluorescent dye are injected under the skin in much the same way as a tattoo. By exposing the tattoo to a light source, the level of fluorescence is indicative of the glucose concentration under the skin in the interstitial fluid.

A glucose biosensor is mainly used to detect the level of glucose concentration in the blood and consists of a biological component coupled to a transducer. The biological sensing is normally performed using either optical or electrochemical methods to generate an electrical signal which is transformed by the transducer into a measurable quantity indicative of the concentration of glucose in the blood. The electrochemical sensing method usually involves a chemical reaction involving a catalyst, whereas optical sensing relies on the absorption properties of glucose to a particular wavelength of light such as infrared. The first glucose enzyme electrodes were developed by Clark and Lyons (1962).

4.2 **AMPEROMETRIC GLUCOSE SENSOR**

Amperometric glucose sensors are enzymatic electrochemical sensors where the measurement process relies on the catalytic oxidation of glucose using an enzyme to produce hydrogen peroxide. The subsequent electrochemical reduction of hydrogen peroxide results in a measurable current in proportion to the concentration of glucose in the blood sample. The uptake of oxygen by glucose is also indirectly proportional to the glucose concentration so the measurement of the relative remaining oxygen with current sensitive electrodes can also provide an indication of the glucose concentration. The most commonly used enzyme is glucose oxidase, although other enzymes such as glucose dehydrase and horseradish peroxidise have been used.

4.2.1 **Glucose Detector Based on Measurement of Hydrogen Peroxide**

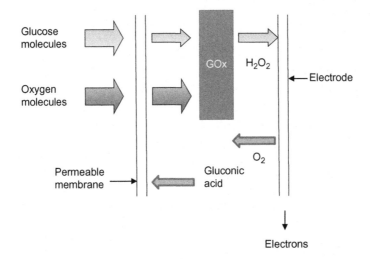

■ **FIGURE 4.1** Enzyme-based amperometric glucose sensor.

Figure 4.1 shows the principle of an amperometric sensor which has an outer permeable membrane to allow oxygen to diffuse through toward the glucose sensor. By making the outer membrane highly permeable to oxygen and less permeable to glucose, any glucose diffusing through the membrane to the enzyme layer is immediately oxidized by the enzyme. Subsequently, the rate of oxidation is determined by the glucose concentration gradient across the membrane. The surface of the working electrode is coated with the glucose oxidase enzyme that acts as a catalyst to convert glucose into gluconic acid which passes back into the bloodstream.

$$\text{Glucose} + \text{GOx}_{(ox)} \rightarrow \text{Gluconic acid} + \text{GOx}_{(red)} \tag{4.1}$$

The reduced enzyme is then oxidized by reacting with oxygen to form hydrogen peroxide.

$$\text{GOx}_{(red)} + \text{O}_2 \rightarrow \text{GOx}_{(ox)} + \text{H}_2\text{O}_2 \tag{4.2}$$

Equations (4.1) and (4.2) can be combined as:

$$\text{Glucose} + \text{O}_2 \xrightarrow{\text{GOx}} \text{Gluconic acid} + \text{H}_2\text{O}_2 \tag{4.3}$$

Using working electrodes made of platinum, biased at the standard electrode potential of approximately 0.7 V, the hydrogen peroxide will be electrochemically oxidized giving rise to a measurable current which is proportional to the concentration of glucose.

$$\text{H}_2\text{O}_2 \rightarrow 2\text{H}^+ + \text{O}_2 + 2\text{e}^- \tag{4.4}$$

The oxidized enzyme will then be reduced and subsequently regenerated by the oxygen to complete the catalytic reduction−oxidation (redox) cycle. However, enzymes are proteins which are soluble in water. Therefore, with the enzyme being mixed in with the reactants and products, the enzyme will start to dissolve or diffuse out of the reaction, leading to a decrease in its catalytic activity. By immobilizing the enzyme in an insoluble rigid structure known as a polymer matrix, the enzyme is effectively separated from the reaction mixture such that the catalytic reaction occurs within the immobilized enzyme layer. This not only helps with the reusability of the enzyme but also improves the sensitivity, linearity, and stability properties of the sensor. Incorporating an immobilized enzyme layer is also known as a "wired-enzyme" system. Techniques mainly used to immobilize enzymes include:

Adsorption—a physical process where dissolved molecules adhere to a solid surface with weak bonding.

Entrapment—a physical process whereby enzymes are trapped in a polymerized porous gel lattice or matrix also known as a sol-gel or composite hydrogel membrane.

Covalent bonding—a chemical process which introduces a functional group and supports matrix with strong bonds.

Cross-linking—a chemical process that can use adsorption or entrapment to covalently bind enzymes to each other to form a polymerized enzyme without a support or enzyme molecules can be cross-linked to form covalent bonds to solid supports.

Another entrapment method widely used relies on the electrochemical immobilization of enzymes in conducting polymers such as polypyrrole (ppy) and polyaniline. The conducting polymer is fabricated in a one-step electropolymerization process where the conducting polymer is electrochemically deposited on the surface of the sensing electrode with the enzyme trapped in the film. Conducting polymers offer good stability, good conductivity, a relatively large surface area, and are easily oxidized.

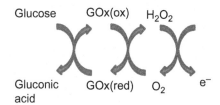

■ **FIGURE 4.2** Enzyme catalytic cycle for oxidation of glucose.

Figure 4.2 shows the enzyme catalytic cycle where the enzyme catalytic reaction occurs in the immobilized conducting polymer layer producing hydrogen peroxide. The Pt/ppy electrodes act as a catalyst to electrochemically oxidize the hydrogen peroxide to produce electron flow in the sensing electrode.

4.2.2 Glucose Detector Based on Measurement of Oxygen

The measurement of oxygen in effect requires two sensors, one with an enzyme-coated electrode to generate a current relating to the uptake of oxygen and the second sensor with a reference electrode (no enzyme) to effectively measure the background level of oxygen. The difference between the two electrode current measurements gives an indication of the glucose concentration.

Oxygen electrodes coated with the glucose oxidase enzyme promote the catalytic reaction between oxygen and glucose:

$$\text{Glucose} + O_2 \xrightarrow{\text{GOx}} \text{Gluconic acid} + H_2O_2$$

Biasing a platinum working electrode negative (0.6−0.7 V) with respect to the reference electrode, the oxygen is reduced resulting in a measurable flow of current:

$$O_2 + 4H^+ + 4e^- \rightarrow 2H_2O \tag{4.5}$$

In all types of sensors that utilize GOx, exposure of glucose oxidase to hydrogen peroxide results in a deactivation of the glucose oxidase enzyme, reducing the sensitivity of the sensor. Because hydrogen peroxide is not utilized for electrochemical measurement in the oxygen detecting type of sensor, means may be employed to reduce the hydrogen peroxide levels in the vicinity of the enzyme. By introducing the catalase enzyme, CAT, this will catalyze the conversion of hydrogen peroxide to water and oxygen minimizing the exposure of GOx to hydrogen peroxide and increasing sensor stability such that:

$$2H_2O_2 \xrightarrow{\text{CAT}} 2H_2O + O_2 \tag{4.6}$$

Both the GOx and CAT enzymes are co-immobilized.

4.3 POTENTIOSTAT MEASUREMENT OF GLUCOSE

A potentiostat circuit shown in Figure 4.3 is commonly used to set up the working electrode voltage and measure the resultant current produced in an amperometric glucose biosensor. The potentiostat has three electrodes: counter electrode (CE), reference electrode (RE) and working

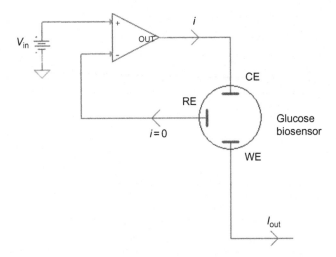

■ **FIGURE 4.3** Amperometric measurement of glucose using a potentiostat circuit.

electrode (WE), also known as the sensing electrode. The amplifier has a large gain and high input impedance such that virtually no current flows into the input pins and the potential difference between the input pins is virtually 0 V. A voltage, V_{in}, applied to the " + " pin will result in an output current flowing through the cell from CE to WE such that the potential difference developed across the reference and working electrodes is the same as that applied to the " + " pin, thus maintaining a virtual 0 V difference across the input pins.

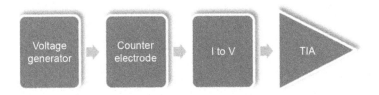

■ **FIGURE 4.4** Block diagram of measurement sensor system.

Figure 4.4 shows a block diagram of a typical system to measure the electrode current. The voltage generator generates a square or pulse waveform which is applied to the CE to set up a potential difference between the CE and the RE. The current to voltage conversion is normally achieved by using a transimpedance amplifier (TIA).

4.4 **NEXT GENERATION OF GLUCOSE SENSORS**

The first generation of glucose sensors use oxygen as a mediator by acting as an electron acceptor to transport electrons from the active enzyme
sites to the electrode surface, generating an electric current. In the second
generation of glucose sensors, a redox reagent acts as the mediator, trans-
porting electrons to the electrode surface.

Other glucose sensors include bioimpedance and optical glucose sensors. Bioimpedance glucose sensors rely on monitoring the impedance
changes of proteins and enzymes. The net charge or dipole that some
proteins exhibit, changes when those protein molecules (target molecules) bind to receptor molecules on the surface of implanted electrodes. Optical glucose sensors determine the concentration of glucose in
the blood by using near-infrared (NIR) spectroscopy which relies upon
the optical absorption properties of glucose in blood. Laser diodes emit
light in the NIR region which is passed through the blood sample. The
spectra of light not absorbed by the blood are picked up by photodiode
detectors which give an indication of the concentration of glucose in the
blood. A decrease in the level of light received together with an
increased light absorption at specific wavelengths indicates an increase
in the concentration of glucose in the blood. The relationship between
NIR blood absorbance and glucose concentration is nonlinear.

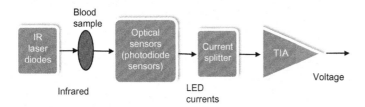

■ **FIGURE 4.5** Block diagram of optical glucose sensor.

Figure 4.5 shows a block diagram of an optical glucose sensor where the
output of the optical sensor consists of a number of Light Emitting Diodes
(LEDs) currents, each representing a specific wavelength of NIR. These currents are passed through a current splitter before being converted into a voltage by using a simple resistor or a TIA, which has the added benefit of
providing a controlled gain. The output voltage represents information on
which NIR wavelengths have passed through the blood sample and, more
importantly, on which wavelengths have been absorbed due to the concentration of glucose in the blood.

Another method to measure glucose concentration in the blood is to place a sensor inside the large-diameter blood vessel in the right atrium junction of the heart (Medical Research Group Inc.) with minimal risk of blood clots or disturbance of the regular atrial flow of blood, which can give a more direct and faster measurement.

Nanosensors provide an alternative technique to implantable glucose detection systems whereby nanosensors are injected into the skin rather like a tattoo. An increase in the blood-glucose level is detected by the tattoo fluorescing under infrared light, indicating a possible lack of insulin (Charles Stark Draper Laboratory, Inc.).

4.5 IMPLANTABLE GLUCOSE SENSOR BY GLYSENS

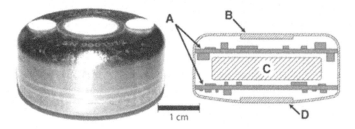

■ **FIGURE 4.6** Implantable GlySens glucose sensor. Cross-sectional view shows electronics module (A), telemetry transmission portal (B), battery (C), and sensor array (D) (Gough et al., 2010). *(Reprinted with permission.)*

The implantable glucose sensor from GlySens is a long-term implant lasting for 1 year or more and does not require continuous calibrations. The implant consists of an integrated glucose sensor with signal conditioning circuits, a wireless telemetry circuit, and a 1-year lifetime battery, all housed in a hermetically sealed titanium housing (Figure 4.6). The wireless radio frequency (RF) link communicates with an external receiver providing continuous glucose monitoring.

The glucose sensor is an amperometric glucose sensor based on the detection of oxygen. The oxygen sensor incorporates dual-enzyme electrode technology with both enzymes, glucose oxidase and catalase, immobilized in a cross-linked protein gel. The catalase enzyme reduces the deactivation of the glucose oxidase enzyme in the presence of hydrogen peroxide, increasing sensor stability and effective lifetime. The reference oxygen sensor contains no enzymes.

An integrated three electrode potentiostatic circuit is used to set up the working electrode voltages and to measure the differential currents between the two oxygen sensors. The sensor array consists of four working-counter platinum electrode pairs and an Ag/AgCl reference electrode, microfabricated onto the surface of an aluminum disk measuring 12 mm in diameter. The enzymes are immobilized by cross-linking with albumin using glutaraldehyde into a gel and are covered with a protective semipermeable membrane layer of polydimethylsiloxane, reducing interference from unwanted molecules.

4.6 IMPLANTABLE CONTINUOUS GLUCOSE MONITORING GLYSENS

■ **FIGURE 4.7** The ICGM system from GlySens. *(Copyright © GlySens. Reprinted with permission.)*

The next generation of glucose sensor from GlySens is the Implantable Continuous Glucose Monitoring (ICGM) System (Figure 4.7), that consists of an implanted glucose sensor with a wireless RF link to an external receiver and monitor, to provide a continuous glucose measurement. The monitor provides a digital readout and graphical display of the glucose levels alerting the user to hypoglycemic and hyperglycemic glucose levels. The implanted sensor is a long-term implant lasting for 1 year or more and does not require continuous calibrations. The ICGM is based on the glucose sensor described previously (Section 4.5) and contains multiple working-counter platinum electrode pairs and amperometric glucose and oxygen sensors using dual-enzyme (glucose oxidase and catalase) electrode technology.

4.7 GLUCOCHIP POSITIVEID CORPORATION AND RECEPTORS LLC

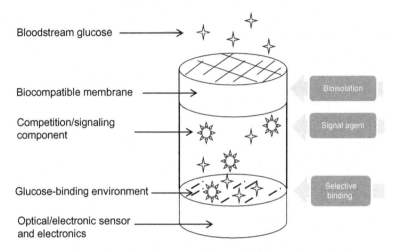

■ **FIGURE 4.8** The GlucoChip glucose biosensor design. *(Copyright © PositiveID Corporation and Receptors LLC. Reprinted with permission.)*

The GlucoChip system design (Figure 4.8) consists of a subcutaneously implanted glucose biosensor embedded onto an RFID microchip which is in contact with interstitial fluid. The biosensor utilizes a synthetic material and a mass sensitive glucose-to-signal transducer to determine the glucose concentration in the blood and to provide a real-time (less than 5 min) glucose number and an indication of the direction of change of glucose concentration. The GlucoChip system will also provide an indication of glucose low, safe, and high limits as defined by the glucose response curve in Figure 4.9. A digital

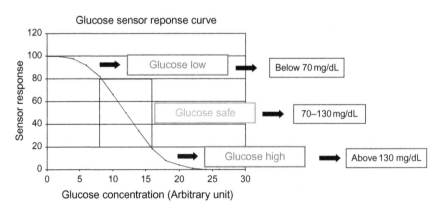

■ **FIGURE 4.9** Glucose response curve with defined limits of glucose levels (Thomas and Carlson, 2011). *(Reprinted with permission.)*

readout of glucose concentration levels as well as patient information such as identity, age, weight, height, etc. can also be provided. An external transponder provides power and data communication to the GlucoChip via an inductively coupled RF link.

The glucose sensor is based upon the principle of a synthetic competitive assay where an analyte (glucose molecule) competes with a competitive binding environment for binding sites on a competitive signaling component.

Figure 4.10 shows the competitive binding environment as an immobilized monosaccharide mimic (iDIOL) and the competitive agent/signaling component as a dendrimer-boronic acid (DBA), all of which compete for the same hydroxyl (iDIOL—OH) binding sites. However, the DBA competitor has a higher affinity for glucose compared to its affinity for the iDIOL, which will result in the net or equilibrium displacement of the DBA molecules from the iDIOL surface.

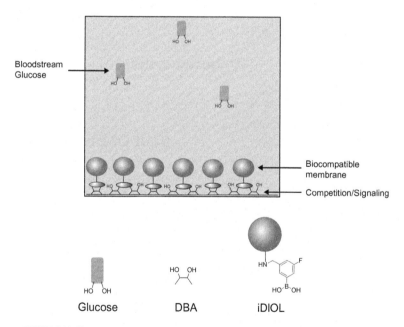

■ **FIGURE 4.10** Three-component competitive binding environment (Thomas et al., 2011). *(Reprinted with permission.)*

As shown in Figure 4.11, when the glucose concentration is low, there is very little competitive interaction between the DBA:iDIOL pairs and the glucose molecules. As the concentration of glucose molecules increases, the DBA molecules, because of their greater affinity for the glucose molecules relative to their iDIOL affinity, are displaced from the DBA:iDIOL

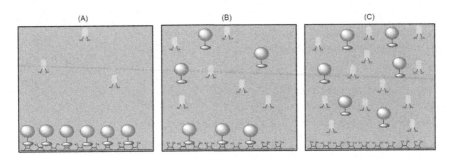

■ **FIGURE 4.11** Indicative glucose interactions: (A) low concentration, (B) safe concentration, and (C) high concentration (Thomas et al., 2011). *(Reprinted with permission.)*

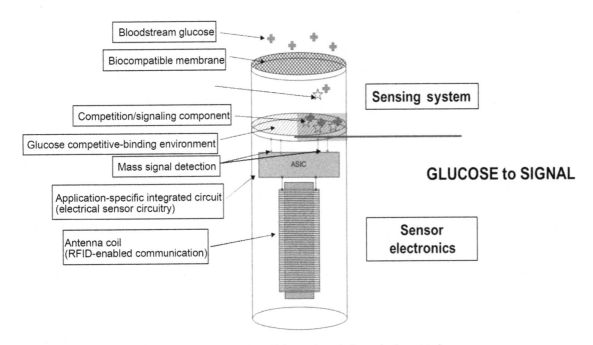

■ **FIGURE 4.12** Components of the GlucoChip glucose biosensor design (Carlson et al., 2007). *(Reprinted with permission.)*

pairs. A MEMS cantilever mechanism based on the detection of resonance frequency and therefore sensitive to the gain or loss of DBA molecules detects the change in mass on the iDIOL surface and generates a proportionate electrical signal which is wirelessly transmitted to the external transponder.

Figure 4.12 shows the construction design of the cylindrical glucose biosensor. The top of the cylinder consists of the sensing chamber which contains the permeable selective membrane and the glucose binding environment. The bottom of the cylinder contains the electronic sensors that detect the flow of generated current in response to the glucose concentration as well as the RFID circuits and antenna coil to wirelessly transmit the data.

BIBLIOGRAPHY

ADA, 2013. American Diabetes Association: Standards of medical care in diabetes 2013 (Position Statement). Diabetes Care 36, S11–S66. Available from: http.//dx.doi.org/doi:10.2337/dc13-S011.

Carlson, R.E., Silverman, S.E., Mejia, Z., 2011. White Paper, Development of the Sensing System for an Implantable Glucose Sensor. Retrieved from PositiveID Corp: http://www.positiveidcorp.com/files/Glucose-Sensor.pdf.

Carlson, R.E., Silverman, S., Meija, Z., 2007. White Paper. Development of an Implantable Glucose Sensor. Retrieved from: http://www.positiveidcorp.com/files/Glucose-Sensor.pdf.

Clark, L.J., Lyons, C., 1962. Electrode systems for continuous monitoring in cardiovascular surgery. Ann. N. Y. Acad. Sci. 102, 29–45.

Gough, D.A., Kumosa, L.S., Routh, T.L., Lin, J.T., Lucisano, J.Y., 2010. Function of an implanted tissue glucose sensor for more than 1 year in animals. Sci. Transl. Med. 2, 42ra53.

IDF, 2007. Guidelines for Management of Postmeal Glucose. International Diabetes Federation (IDF), ISBN 2-930229-48-9. Available from: http://www.idf.org/webdata/docs/Guideline-PMG-final.pdf.

Thomas, C., Roska, R.L.W., Carlson, R.E., 2011. Development of a Synthetic, Closed-Cycle Sensing System for an Implanted Glucose Sensor. 11th Annual Diabetes Technology Meeting.

Thomas, R., Carlson, R., 2011. White Paper. Development of the Sensing System for an Implantable Glucose Sensor. Retrieved from: http://www.positiveidcorp.com/files/Glucose-Sensor.pdf.

WHO, 2006. Definition and Diagnosis of Diabetes Mellitus and Intermediate Hyperglycemia. Report of a WHO/IDF Consultation. World Health Organization, ISBN 92 4 159493 4 (NLM classification: WK 810): ISBN 978 92 4 159493 6.

Chapter

Cochlear Implants

5.1 INTRODUCTION

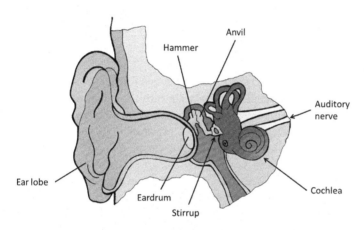

■ **FIGURE 5.1** Structural features of the human ear involved in hearing.

The anatomical structures of the human ear involved in hearing consist of the outer, middle, and inner ear. The ear lobe forms part of the outer ear which helps to "funnel" sound into the middle ear, which consists of three small bones, the hammer (malleus), anvil (incus), and stirrup (stapes), all mechanically coupled together (Figure 5.1). A sound pressure displacement of the ear drum pushes against the hammer, which taps against the anvil, which causes the stirrup to push against the oval window of the inner ear, all in sympathy with the external sound. The physical coupling acts like a lever which effectively amplifies small noise levels, turning variations in sound pressure into mechanical vibrations in the middle air. The inner ear, also known as the cochlea, is a spiral "snail"-shaped bone cavity structure consisting of just over two turns about its axis and is filled with fluid called perilymph. The mechanical vibrations from the stirrup acting on the oval window sets up traveling sound waves in the fluid of the cochlea or inner ear.

Implantable Electronic Medical Devices. DOI: http://dx.doi.org/10.1016/B978-0-12-416556-4.00005-X

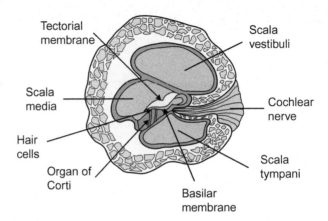

Tectorial
membrane

Scala
vestibuli

Scala
media

Cochlear
nerve

Hair
cells

Organ of
Corti

Scala
tympani

Basilar
membrane

■ **FIGURE 5.2** Cross section of the cochlea.

A cross section of the cochlea reveals three chambers or ducts, the scala vestibuli, scala tympani, and the scala media, as shown in Figure 5.2. Between the scala tympani and the scala media is the basilar membrane on which the structure known as the organ of Corti sits. Four rows of sensory hair cells on the organ of Corti vibrate from the traveling waves set up in the fluid and subsequently along the basilar membrane. These hair cells are comprised of stereocilia that are attached to the tectorial membrane such that the tectorial membrane moves back and forward in response to sound waves. The inner row of hair cells bends in response to the maxima and minima of the traveling waves, resulting in ionic currents flowing in and out of their respective cell membranes, generating action potentials in the auditory nerve fibers. The outer three sensory rows of hair cells are occupied by the outer hair cells, which do not stimulate the auditory nerves but undergo a shape change in response to traveling sound waves. They are thought to contribute to the selective tuning of frequencies at points along the basilar membrane by mechanically amplifying low-intensity sound levels.

The basilar membrane that runs through the cochlea has a structure that is initially narrow and thick at its base at the oval window nearest to the middle ear, and then decreases in width but increases in thickness, becoming more rigid at the apex which is at the center of the spiral coil structure. A received sound wave in the outer ear is transduced into mechanical vibrations in the middle ear, which sets up a pressure wave in the fluid of the cochlea, causing the basilar membrane to vibrate, setting up resonating standing waves along its length where the amplitude of displacement along the membrane is dependent on the received audio frequency.

Due to the anatomical structure and shape of the membrane, different sections of the membrane respond to different frequencies. The wide, thin end of the membrane at the apex end vibrates at maximum amplitude when subjected to low-frequency standing waves, whereas the narrow, thick basal end resonates at higher frequencies. Overall, the membrane structure gives rise to a frequency-selective membrane such that the membrane effectively decomposes the incoming sound wave into its frequency components. The frequency response of the membrane is exponential in nature which is also known as a tonotopic response. Figure 5.3 shows the tonotopical arrangement of the cochlea, which shows a high to low frequency response toward the apical end of the basilar membrane.

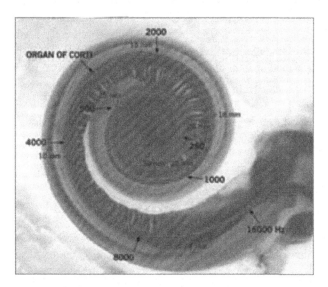

■ **FIGURE 5.3** Tonotopical arrangement of the cochlea. *(Copyright © MED-EL. Reprinted with permission.)*

5.2 **TYPES OF HEARING LOSS**

Hearing loss can be conductive, sensorineural, or a combination of both. Conductive loss occurs when sounds are not able to pass from the outer to the inner ear and with age (presbycusis), the transmission of sound waves from the outer to the inner ear diminishes due to factors such as the eardrum losing some of its elasticity and the bones in the middle ear losing some mobility, becoming more rigid in their action. One common cause is known as otosclerosis, in which abnormal bone growth around the base of the stapes becomes thicker, restricting the movement of the stapes.

Sensorineural deafness is mainly related to inner ear damage of the sensory hair cells or auditory nerve damage. These damages can be age related or induced by conditions relating to infection such as meningitis, congenital malformations, or acoustic trauma due to prolonged exposure to loud noises. Very loud noises generate extreme shear forces on the tectorial membrane, resulting in partial or permanent damage to the hair cells, which cannot be replaced.

5.3 **COCHLEAR IMPLANTS**

The cochlear implant was developed to help people who have partial or profound deafness resulting from damage due to injury or a condition which may be age related or induced. The human ear consists of the outer, middle, and inner ear which together effectively couple and translate sound waves into electrical stimuli for the brain to process and perceive as sound. It is the vibration of sensory hair cells in the inner ear, in response to sound waves, that results in the generation of electrical stimuli in the auditory nerve. However, if the sensory hair cells become damaged or lose their function, this can lead to deafness. The cochlear implant effectively bypasses the sensory hair cells by inserting an electrode array into the inner ear and directly stimulating the auditory nerve in response to an audio signal from a microphone placed external to the ear.

5.4 **COCHLEA ELECTRODE ARRAYS**

Electrodes inserted into the cochlea are designed to be atraumatic, causing minimal trauma to the delicate structure of the cochlear on insertion, as the size and shape of the electrode array directly affects the incidence of observed trauma (Rebscher et al., 2008). The risk of trauma potentially increases with electrode insertion depth required to increase the range of neurons that can be stimulated by accessing the low-frequency neurons located toward the apex of the scala tympani. The depth of insertion is measured as the number of cochlear turns or the angle of rotation measured from the beginning of the base of the cochlear at the round window, to the most apically inserted electrode contact.

Electrode insertion is normally made through the round window, although in some cases, a cochleostomy is performed whereby the electrode is inserted through a small hole made in the basal turn of the cochlear. In either approach, the inserted electrode is designed to follow the curved channel of the scala tympani. Straight electrodes, also known as lateral electrodes, tend to follow the outer wall curvature of the cochlea, whereas preformed curved perimodiolar electrodes are designed to follow the inner wall curvature of the cochlea, where the neural cells are predominantly located. Placing the electrode contacts in close proximity to the neural cells can effectively reduce stimulation threshold currents and the spread of electric field potentials between electrodes, resulting in a potential increase in discrete frequency perception.

5.5 **SPEECH CODING**

Speech processing is dependent on the perceived pitch of sound related to regions of the basilar membrane that elicit a neural response in the auditory nerve which is transmitted to the higher centers in the brain, where it is decoded and perceived as sound. Consequently, speech processing can therefore be separated into spatial and temporal encoding of generated electrical signals in the auditory nerve. The spatial encoding is provided by the relative positioning of electrodes placed along the basilar membrane and the temporal encoding by the relative timing of stimulation pulses or rate of stimulation. The greater the number of electrodes, the finer the frequency resolution that can be processed. Electrodes placed near the basal end of the membrane will be stimulated in response to high-frequency acoustic sounds, whereas electrodes placed near the apex of the membrane will be stimulated in response to low-frequency acoustic sounds.

Human hearing has a perception of sound ranging from the faintest sound heard defined as 0 dB to soft whispers typically at 20 dB to loud rock music at 140 dB. This gives the ear an input dynamic range (IDR) of 120 dB. In comparison, normal speech is considered to lie in the range of 40−60 dB. In order to qualify for a cochlear implantable device, one must be profoundly deaf in both ears with a measured hearing loss of 90 dB or more.

5.6 **COCHLEAR IMPLANT SYSTEMS**

Modern cochlear implant systems consist of an external microphone, a speech processor, an RF data transmitter, internal receiver and stimulator, and an electrode array which is surgically inserted through a small incision made at the back of the ear, into the cochlea through either the round or oval window. The electrode array consists of fine wires of different lengths bundled together and inserted into the scala tympani canal such that the exposed wire tips will end up being positioned alongside the basilar membrane at set distances relative to each other.

In order to map individual electrodes to specific frequencies, modern cochlear implants use speech-coding strategies to separate the external speech signal into frequency bands which are analyzed for their spectral content to amplitude-modulate a biphasic pulse train that directly or indirectly activates the stimulating electrode. Figure 5.4 shows some of the principal processing blocks used in many speech-coding strategies.

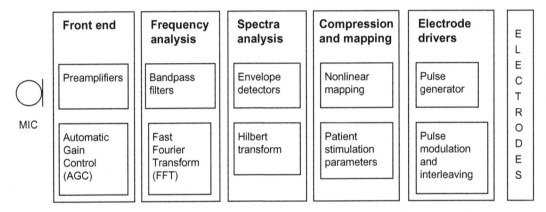

■ **FIGURE 5.4** Principal processing blocks used in many speech-coding strategies.

In speech, the higher frequencies provide intelligibility for speech but they have a low sound intensity compared to the low-frequency components in speech. The characteristics of the microphone circuits used in cochlear implantable systems are such that they exhibit a higher gain at higher frequencies compared to low frequencies. This then provides a preemphasis of high-frequency components of the audio signal.

The Automatic Gain Control (AGC) maintains the sound level of speech to within predetermined limits and is usually based upon a compression amplifier that reduces the gain (compression) when the sound intensity increases, thus preserving the signal envelope and spectral content.

The speech signal is separated into frequency bands also known as channels, using filters or by applying the fast Fourier transform (FFT) to the signal. The envelope detector determines the amplitude of the signal and is usually performed using either a diode rectifier followed by a low pass filter or by applying the Hilbert transform to the filtered bandpass signal.

The dynamic range of stimulation is dependent on the patient's measured threshold levels between a barely detectable audible level and that of a maximum comfortable level of a loud sound. Subsequently, the envelope detector outputs are compressed to map to the patient's implant dynamic range. The compression function is nonlinear and is usually implemented using logarithmic or power-law functions.

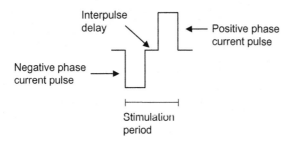

Negative phase current pulse

Interpulse delay

Positive phase current pulse

Stimulation period

■ FIGURE 5.5 Biphasic stimulation pulse.

The stimulation pulse consists of biphasic pulses as shown in Figure 5.5 where the negative going pulse which initiates a neural response is charge balanced by the positive going pulse. The charge interpulse delay helps to minimize the effect of the negative going-evoked reaction by introducing a suitable delay between both phases of the pulse train. However, simultaneous electrode stimulation will result in an interaction of electric fields between adjacent electrodes, resulting in unwanted current flow between electrodes that can distort or reduce the temporal spectral resolution of the cochlear implant. Consequently, a number of speech-coding techniques such as Continuous Interleaved Sampling (CIS), Advanced Combinational Encoder (ACE), Spectral Peak Extraction (SPEAK), and HiRes have been developed to minimize this effect.

The stimulus from an individual electrode will evoke a response pertaining to a perceived pitch at the electrode site on the basilar membrane. If two adjacent electrodes are stimulated, the perceived pitch will lie somewhere between the two pitches perceived with two individual electrodes, thus creating a virtual channel. By varying the ratio of the magnitude of current between a pair of electrodes, or by varying the ratio of the stimulation rate between a pair of electrodes, an intermediate pitch can be perceived between the sites of the electrode pair. In effect, the number of stimulation sites can be increased with virtual channels rather than increasing the number of electrodes, which has practical limitations.

5.7 **CONTINUOUS INTERLEAVED SAMPLING**

Continuous interleaved sampling (CIS) reduces the problem of current interaction between adjacent electrodes by maintaining a constant stimulation rate for each channel and interleaving the biphasic stimulation pulses. The trains of nonoverlapping biphasic pulses stimulate each electrode in turn, minimizing the interaction of overlapping electric fields and hence current interaction between adjacent electrodes.

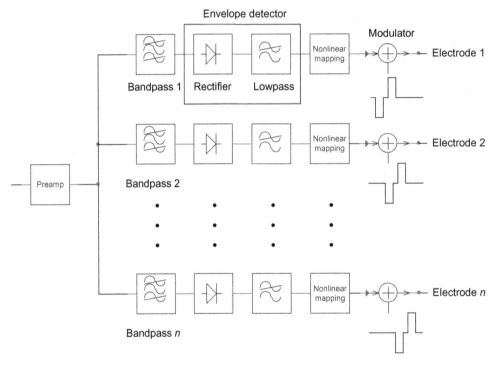

■ **FIGURE 5.6** A CIS-based cochlear implant system.

Figure 5.6 shows a block diagram of a CIS system in which the speech signal is band-pass filtered into a number of subbands and then each subband signal is rectified and low pass filtered to extract the temporal envelope. The nonlinear mapping block is customized to the user's premeasured dynamic range and the resulting temporal envelope is used to amplitude-modulate an interleaved biphasic pulse train at a constant rate, which then becomes the applied stimulus to the auditory nerve. A typical stimulation rate for CIS is 1500 pulses/s per channel. A variant of CIS includes HiRes, which uses high pulse rates typically between 1500 and 1900 pulses/s per channel.

5.8 **HIRES120**

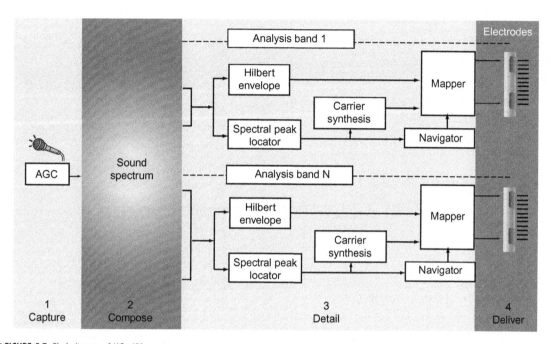

■ **FIGURE 5.7** Block diagram of HiRes120.

This is a software algorithm that uses current steering to create a maximum of 120 stimulation sites using a 16-electrode array, resulting in 15 electrode pairs. There are 8 proportional current ratios that can be set between each electrode pair, giving a total of 120 virtual channels. Figure 5.7 shows the block diagram of the HiRes120 speech-coding strategy.

The sound received from the external microphone is sampled using an A/D converter to produce a digital audio stream that is converted into the frequency domain using a 256-point FFT before being separated into 15 individual channels. With HiRes120, temporal and spectral analysis is performed in parallel. Temporal analysis is performed using the Hilbert transform, while the Navigator determines the spectral maximum amplitude for each electrode pair. The frequency at which the spectral maximum occurs is then used to determine the rate of stimulation and the relative current ratios between electrode pairs to effectively steer the stimulus current to create a virtual channel in between electrode pairs.

5.9 LIFESTYLE™ COCHLEAR IMPLANT SYSTEMS BY ADVANCED BIONICS™

(A)

(B)

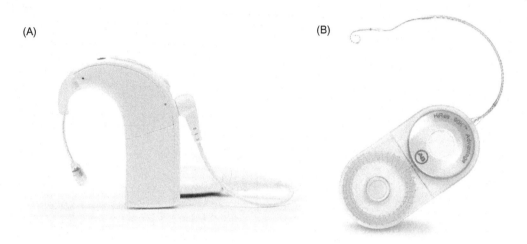

■ **FIGURE 5.8** The lifestyle™ cochlear implant system consisting of (A) the Naída CI Q70 Sound Processor and (B) the HiRes90K cochlear implant. *(Copyright © Advanced Bionics. Reprinted with permission.)*

Figure 5.8 shows the lifestyle™ cochlear implant system consisting of the Naída CI Q70 Sound Processor and the HiRes90K cochlear implant. The processor can use the HiRes120 and its modified power-saving version, the HiRes Optima speech processing strategies. The Neptune sound processor is a waterproof version that can be used for swimming and other water activities. The sound processors can be controlled wirelessly and the AccessLine™ T range of accessories for wireless connection to Bluetooth, MP3, and other audio devices is available for the Naída CI Q70™.

5.10 CLEARVOICE™

ClearVoice™ from Advanced Bionics is a software enhancement algorithm that can be applied to the Advanced Bionics Naída CI Q70™ and Neptune™ sound processors in conjunction with the HiRes Fidelity 120™ sound-processing strategy. By identifying and comparing noise levels and subsequent signal-to-noise ratio (SNRs), the channel gain is dynamically reduced for nonspeech sounds and increased for low-level speech in order to enhance speech discrimination.

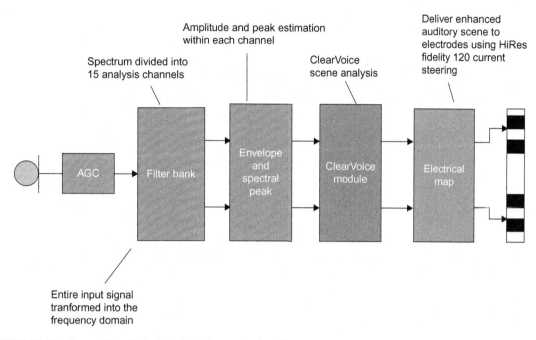

■ **FIGURE 5.9** Block diagram of HiRes Fidelity 120 and ClearVoice signal analysis. *(Copyright © Advanced Bionics. Reprinted with permission.)*

In Figure 5.9, the sound signal from the microphone front end is digitized and frequency analyzed using a 256-point FFT. The envelope and spectral peaks, extracted using a Hilbert transform, are analyzed to determine which electrode pairs are to be used for current steering. At the same time, a signal-to-noise estimator analyzes and classifies the incoming sound signal in order to reduce background noise from relevant sound information such as speech. For example, a channel with a similar energy level to that of estimated background noise will be attenuated, whereas a comparably higher channel energy level will be sustained for stimulation. The amount of channel attenuation is determined by comparing the difference between the incoming energy content of a sound signal and the output of the signal-to-noise estimator. The amount of attenuation provided by ClearVoice™ can be programmed, providing a maximum of 6, 12, or 18 dB levels of attenuation.

The Advanced Bionics family of 16-electrode arrays includes the HiFocus™ Mid-Scala (Figure 5.10), which has a formed precurved electrode array structure and is designed to "free float" in the scala tympani in order to reduce the risk of damage to the delicate structures of the cochlear upon insertion through the round window. The HiFocus™ 1j has a slightly curved

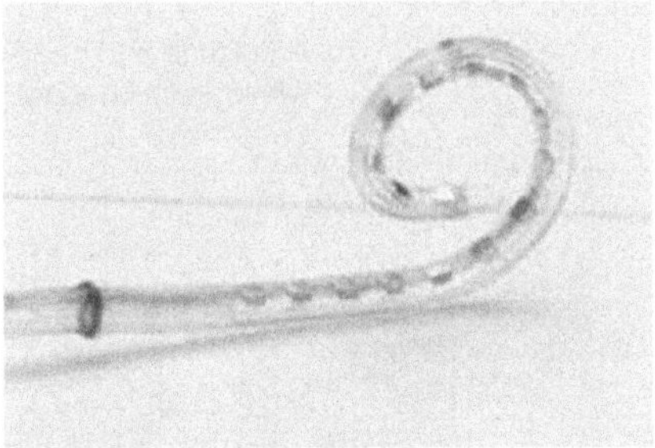

■ **FIGURE 5.10** HiFocus Mid-Scala electrode array. *(Copyright © Advanced Bionics. Reprinted with permission.)*

electrode array with raised partitions between electrodes designed to reduce the electric field interactions between electrodes. The perimodiolar Focus™ Helix electrode is designed to conform to the cochlea's natural contour.

5.11 N-OF-M, SPECTRAL PEAK EXTRACTION (SPEAK) AND ADVANCED COMBINATIONAL ENCODER (ACE)

In comparison to CIS, in which all available channels are stimulated, the n-of-m, SPEAK, and ACE sound-coding strategies select a reduced number of frequency bands for stimulation in order to increase the contribution of the temporal resolution relative to the spectral content of the acoustic signal. Only a subset of the total number of channels that contain the most dominant spectral information are selected. This inherently reduces the intrusion of less-dominant sounds and reduces the density of neural stimulation.

In n-of-m, the audio signals are filtered into "m" frequency bands, otherwise known as channels, which are representative of the number of physical electrodes in the implantable system. The signal envelopes are then analyzed to identify "n" signals, which have the highest amplitudes (spectral maxima). These n-of-m channels are then selected to deliver a stimulus pulse to the corresponding electrode. For example, for a 6-of-12 strategy, only 6 out of the 12 available channels, corresponding to the highest signal amplitudes, will be selected within each time window of analysis.

For n-of-m and ACE, the number of channels with the highest signal amplitudes, "n," is fixed. Although SPEAK is broadly similar, there is a slightly greater dependence of the number of selected channels on the stimulation levels used, causing a somewhat more adaptive value of "n." SPEAK separates the signal into 20 frequency bands and typically selects between 6 and 10 channels that have the highest energy content relating to the maximum spectral amplitude. SPEAK stimulates at a channel rate of approximately 250 Hz, whereas ACE has a higher stimulation rate ranging from 250 Hz to 3.5 kHz per channel with a default rate of 900 Hz per channel.

Increasing the number of stimulation electrodes results in an increase in the spread of stimulation currents into the cochlea, which can interfere with the resolution of the frequency that can be perceived by the user. By decreasing the number of stimulation electrodes, the stimulation current is more localized, which can give a better spatial frequency resolution. Consequently, only a subset of the total number of channels which contain the most dominant spectral information are selected. The total stimulation rate is given as the product of the channel rate and the number "n" of selected channels.

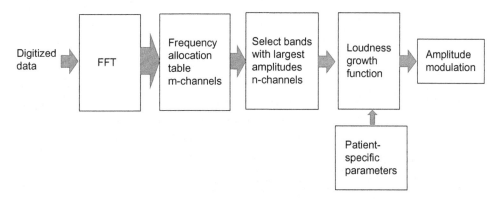

■ **FIGURE 5.11** Block diagram of ACE sound-coding strategy.

Figure 5.11 shows the block diagram of a typical ACE strategy with an 8 kHz audio signal sampled at 16 kHz. The sampled data is initially multiplied by a 128-point Hann Window Function to reduce boundary edge effects introduced by the FFT. With an audio sampling frequency of 16 kHz, the period of the FFT sample window is given by:

$$\frac{\text{No. of samples}}{\text{Audio sampling rate}} = \frac{128}{16\,\text{kHz}} = 8\,\text{ms}$$

which separates the acoustic signal into 64 individual frequency amplitudes spaced 125 Hz apart. The frequency band outputs are then combined to create a frequency allocation table with the same number of channel outputs as there are available electrodes.

The table can be customized to provide the required linear-logarithmic frequency spacing characteristic required to enhance the characteristic frequency resolution of speech such as vowel recognition. Typically, the ACE filter bank is designed with a linear frequency up to 1 kHz and a logarithmic spacing above 1 kHz. The frequency bands with the highest spectral energy content (spectral maxima) are then used to select the corresponding electrodes for stimulation. The loudness growth function maps the user's premeasured dynamic range parameters and resulting temporal envelope to amplitude modulate an interleaved biphasic pulse train as in the case of the CIS strategy, which is then applied to the appropriate electrode.

The calculated sample window period of 8 ms does not necessarily dictate the stimulation rate for an individual channel because the sample window can be shifted by a smaller increment than 8 ms to deliver whatever stimulation rate is required. However, the shifting analysis window inherently introduces some temporal filtering, as a proportion of the same audio samples are used to calculate the next energy estimate.

The total stimulation rate is given as:

$$\text{Total stimulation rate} = \text{Channel stimulation rate} \times \text{"}n\text{" selected channels}$$

The number of selected channels "n" can be increased, resulting in an increase in the spectral representation of stimulation, but as the total stimulation rate for an implant has an upper limit (by design), the channel stimulation rate may need to be decreased if a high "n" is desired. By keeping the number of channels at half the number of available channels, the stimulation current is more localized, giving an improved frequency perception at the expense of reducing some temporal content. Consequently, an inherent balance exists between the number of selected channels and the stimulation rate for optimal performance, especially as the number of available channels increases.

5.12 NUCLEUS® 6 SYSTEM, COCHLEAR

The Nucleus® 6 System (Figure 5.12) consists of the CP910 and the smaller CP920 sound processors. Both processors support wireless connectivity of audio accessories such as the Mini Mic, TV Streamer, and Phone Clip+ (which incorporates Bluetooth connectivity) identical to those from GN Resound. Both sound processors also have an inbuilt telecoil for use with induction loops. Only the CP910 has an accessory port for wired connection to audio accessories such as MP3 players. The Nucleus® 6 System also has the option of a Remote Control CR210 and

■ **FIGURE 5.12** Nucleus® 6 System showing CR210 remote control, the CR230 remote assistant, and the CP922 sound processor. *(Copyright © Cochlear. Reprinted with permission.)*

Remote Assistant CR230 handheld controller. The CR230 allows for remote monitoring of the cochlear system, manual adjustment of the hearing environment, and diagnostics.

The processors incorporate dual-calibrated microphones which allows for the possibility to automatically adjust and focus on sounds of interest from the front while blocking out unwanted background noise to the sides and rear to effectively make the received sounds clearer. The processors have an optional auditory scene classifier algorithm, SCAN, to analyze and identify the external sound environment and subsequently adjust the microphone properties and other algorithms in order to optimize the perceived clarity of hearing. The six scenes that SCAN can currently identify are Quiet, Speech, Speech in Noise, Noise, Music, and Wind noise.

The default sound-coding strategy for the Nucleus® 6 processors is ACE and is the most widely used strategy. However, the Nucleus® 6 also supports CIS, SPEAK, and MP3000, which is a similar strategy to ACE but where a reduced number of maxima are selected based upon a psychoacoustic model of audibility. The default stimulation rate is 900 Hz with a default number of maxima set at 8. The available stimulation rates are 250, 500, 720, 900, 1200, 1800, 2400, and 3500 Hz. The number of available maxima ranges from 1 to 20, but the total stimulation rate, given as the product of the channel rate and maxima, may not exceed 31,500 Hz.

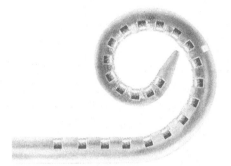

■ **FIGURE 5.13** Contour Advance™ electrode array. *(Copyright © Cochlear. Reprinted with permission.)*

The Nucleus family of 22-electrode arrays consists of the perimodiolar Contour Advance™ electrode (Figure 5.13), which has a preformed spiral shape and also four different lateral wall electrodes. These in turn comprise of the "full-banded" straight electrode intended for many varied cochlea anatomies; the "double array" comprising two separate arrays of 11 platinum electrodes designed to be inserted into both the basal and medial turns of the cochlea when the cochlea has become obstructed by bone growth (ossification); the Hybrid L24, which is a short electrode intended for optimal low-frequency hearing preservation; and the "slim straight" electrode, which balances the insertion depth and residual hearing preservation qualities into a single design.

5.13 **DUAL-LOOP AGC**

The dynamic range of normal hearing is considered to be 100−120 dB, whereas the equivalent IDR for electrical stimulation is in the range of 10−20 dB. However, the cochlear implant not only has to accurately replicate the range and levels of sound associated with speech, but also needs to respond to the sudden transient changes in sound intensity levels that give speech its subtleness and intelligibility. Therefore, the AGC in the front end of the audio processing stage of a cochlear implant must respond quickly in order to detect sudden transient changes in sound intensity.

Slow-responding AGCs can effectively miss sudden transient increases in sound levels, leading to sudden uncomfortable sound levels being conveyed to the user. Too quick a compression of the AGC (attack time) and slow release time can result in an unwanted electrically generated sound level.

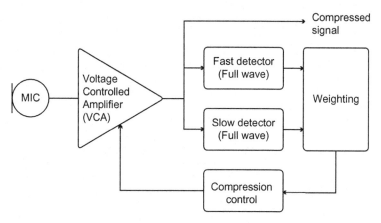

■ **FIGURE 5.14** MED-EL Dual-loop AGC system.

Figure 5.14 shows the dual-loop AGC system from MED-EL where the signal from the microphone is simultaneously passed to two peak detectors connected in parallel. One detector responds to slowly varying signals, while the other detector responds to quickly changing signals. Either detector can adjust the microphone gain in response to quiet acoustic levels being amplified and loud transient signals being compressed. In normal operation, the slow detector is in control for most of the time, but with sudden loud transient signals the fast detector takes control, reducing the gain and hence compressing the components of the loud sound signal to acceptable levels. This results in a fast adaptive gain control system with an IDR of 75 dB over a range of 25–100 dB.

5.14 **FINE STRUCTURE PROCESSING**

As shown in Figure 5.15, an acoustic signal can be decomposed into two components, a slowly varying envelope and a constant amplitude high-frequency carrier wave, much the same as amplitude modulation. The high-frequency signal is referred to as the fine-structure and carries acoustic cues that are considered to be important for pitch, especially when listening to music, whereas the envelope provides sufficient information to understand speech, especially in quiet environments. Consequently, speech-coding strategies using a fixed rate analysis of the sound envelope by themselves will not provide adequate fine-structure clarity required for the perception of music. The interaural time delays

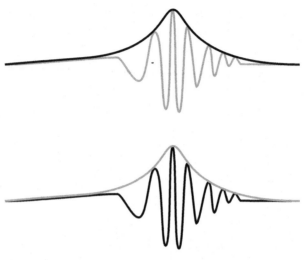

■ **FIGURE 5.15** Envelope and fine-structure components of an acoustic signal. *(Copyright © MED-EL. Reprinted with permission.)*

incorporated in the fine-structure acoustic signals also help to provide information as to the origin of an external sound source.

Place coding relates to the tonotopic arrangement of the cochlea in that specific sites on the basilar membrane respond to specific frequencies with low-frequency responses located at the apical (apex) end and high-frequency responses located at the basal (base) end.

Phase locking, also known as temporal (time) coding, occurs when the sound pressure wave, the resulting standing wave along the basilar membrane and neural activity from the inner hair cells, are all time related. The resultant synchronous neural activity therefore provides for the fine structure of the sound signal. Frequencies up to several kilohertz are place coded and phase locked, whereas with higher frequencies, only place coding is prevalent. Based on this and the inherent technical limitations of sampling and subsequent stimulation rate, the fine structure-processing strategy only provides temporal fine-structure information in addition to envelope information relating to the low-frequency channels, whereas the mid- and high-frequency channels provide envelope information only.

Figure 5.16 shows a block diagram of the FS4 strategy used in the MED-EL OPUS 2 audio processors and the RONDO audio processors, which is a development of the CIS speech processing strategy. In parallel with the envelope detectors for the first two to three channels (200–300 Hz) are zero crossing

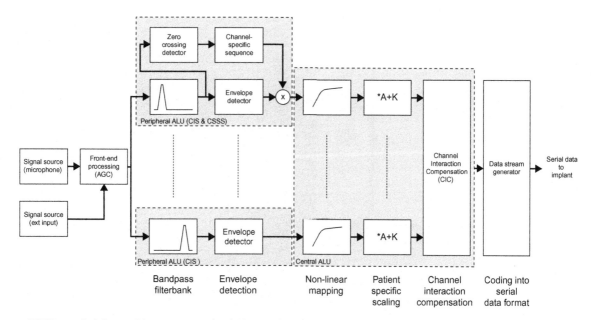

■ **FIGURE 5.16** Block diagram of the FS4 strategy used in the Opus 2 audio processor. *(Copyright © MED-EL. Reprinted with permission.)*

detectors that produce a series of channel-specific sampling sequence (CSSS) pulses such that the instantaneous repetition rate of a CSSS is the same as the instantaneous fine-structure frequency from the respective band-pass filter. Combining the CSSS with the output of the envelope detector gives the fine structure and envelope information for the respective band-pass filter.

The MED-EL FS4 strategy is able to provide temporal fine-structure information on the first four channels by focusing a higher stimulation rate on the first four channels. An additional strategy, FS4-p, uses parallel stimulation to augment the temporal accuracy of fine-structure information which is provided in combination with channel interaction compensation to predict and minimize the real-time spread of electric current between adjacent electrodes.

5.15 MAESTRO™ COCHLEAR IMPLANT SYSTEM BY MED-EL

The MAESTRO™ cochlear implant consists of the OPUS 2 and OPUS 2 XS behind the ear external audio processor and the CONCERTO implant (Figure 5.17). The alternative RONDO single unit device is attached by magnets to the outside of the skull and can be hidden under long hair. Accessories include a Fine Tuner™ remote control to adjust volume and

■ **FIGURE 5.17** MAESTRO™ cochlear implant system consisting of the CONCERTO implant and a choice of either the OPUS 2 XS behind-the-ear or RONDO single-unit audio processors. *(Copyright © Med-El. Reprinted with permission.)*

sensitivity levels and a range of both MED-EL and third-party assistive listening devices to connect to induction loop coils, Bluetooth, MP3 players, and iPod® devices.

There are three families of lateral wall electrode arrays which incorporate electrodes of different lengths; the FLEX, FORM, and CLASSIC series. Within each of these series, a diversity of lengths is offered, varying between 19 and 31.5 mm, to accommodate the intercochlear variability as well as to accommodate the different etiologies. The MED-EL CLASSIC family includes the Standard, which is tapered and designed for atraumatic deep insertion into the apical end of the cochlea for complete cochlear coverage; the Medium for a shorter scala tympani; and the Compressed, where the electrodes are spaced closer together toward the distal apical end to cater for partial ossification.

The FLEX family with electrodes varying between 20 and 31.5 mm in length utilizes proprietary FLEX-Tip™ technology, which features single contacts at the leading end, ultra flexible wave-shaped wires, and a tapered tip for increased mechanical flexibility (Figure 5.18).

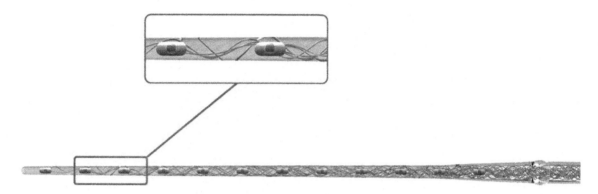

■ **FIGURE 5.18** FLEX electrode. *(Copyright © MED-EL. Reprinted with permission.)*

The FORM electrode arrays, consisting of a 19 and 24 mm version, are designed for malformed cochlea and incorporate an integrated SEAL where there is a risk of cerebrospinal fluid leakage.

BIBLIOGRAPHY

Rebscher, S.J., Hetherington, H., Bonham, B., Wardrop, P., Whinney, D., Leake, P.A., 2008. Considerations for design of future cochlear implant electrode arrays: Electrode array stiffness, size, and depth of insertion. J. Rehab. Res. Dev. 45 (5), 731–748.

Pacemakers and Implantable Cardioverter Defibrillators

6.1 INTRODUCTION

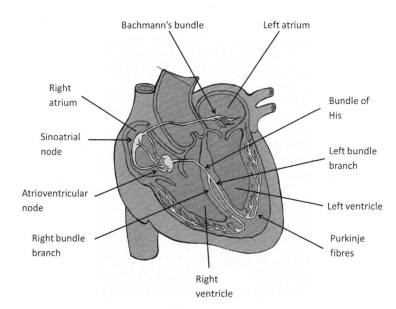

■ FIGURE 6.1 Main anatomical features of the heart.

The heart is a four-chamber muscular organ that receives and pumps blood to the pulmonary and systemic circulatory systems of the human body (see Figure 6.1). The pulmonary system consists of the left atrium chamber which fills with oxygenated blood from the lungs and the right ventricle chamber, from which deoxygenated blood is pumped to the lungs. The systemic system consists of the left ventricle chamber from

Implantable Electronic Medical Devices. DOI: http://dx.doi.org/10.1016/B978-0-12-416556-4.00006-1

which oxygenated blood is pumped to the arteries and capillaries to supply oxygen to cell tissues in the human body and the right atrium, which fills with deoxygenated blood returning from the systemic veins. The tricuspid valve between the right atrium and right ventricle and the mitral valve between the left atrium and left ventricle' prevent the backflow of blood from the ventricles into the atria.

There are two main types of cardiac muscle cells in the heart: pacemaker cells that have no resting cell membrane potential but spontaneously depolarize and nonpacemaker cells that have a resting potential and conduct impulses very quickly when depolarized by an electrical stimulus. The atria and ventricles are made up predominantly of nonpacemaker cells such that a stimulus applied to nonpacemaker cells in the atrium spreads to all the atrium cells, causing both atria to contract. The same applies to the nonpacemaker cells in the ventricles, in which an applied stimulus will spread, resulting in a contraction of both ventricles. Pacemaker cells in the heart are responsible for causing the heart muscles to rhythmically contract and relax at regular intervals; This provides the necessary pumping action to pump blood around the body. One heartbeat corresponds to one cardiac cycle consisting of the systole phase (the atria and ventricles contract) and the diastole phase (the atria and ventricles relax).

The heart's cardiac cycle of rhythmic muscle contractions and relaxations starts with the generation of an electric impulse from a small group of pacemaker cells in the sinoatrial (SA) node located in the upper right lateral wall of the right atrium. The SA depolarization rate determines the nominal heart rate and rhythm, which can vary between 60 and 100 beats per minute (bpm). As the atria start to fill with blood, the pacemaker cells are spontaneously depolarizing, causing action potentials to spread to other myocardial cells in the right atrium and via Bachmann's bundle to the left atrium. The action potentials from the SA node simultaneously propagate down via a conductive neural path toward the atrioventricular (AV) node which consists of cells that depolarize at a slower rate (50 bpm) compared to the pacemaker cells in the SA node. Consequently, the conduction of the impulse from the SA node slows down, introducing sufficient delay for complete contraction of the atria before the ventricles start to contract. The impulses then continue to propagate down through the bundle of His and the Purkinje fibers and the left and right bundle branches, resulting in simultaneous contraction of the ventricles and ejection of the blood from the ventricles. The subsequent electrical activity of the heart muscles can be detected as an electrocardiogram (ECG) electrical signal, as shown in Figure 6.2.

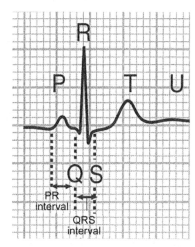

■ **FIGURE 6.2** ECG waveform.

With reference to Figure 6.2:

P wave represents the depolarization of the atrial pacemaker cells as the action potential propagation travels down from the SA node toward the AV node. Typical propagation or conduction times are between 80 and 100 ms.

PR interval, measured from the onset of the P wave to the beginning of the QRS complex, represents the time delay between atrial depolarization and ventricular depolarization. Typical values range from 120 to 200 ms.

QRS complex represents ventricular depolarization. Typical values range from 60 to 100 ms.

R wave represents the start of ventricular contraction and is known as the "early ventricular depolarization."

S-T segment corresponds to the time the ventricles are depolarized. Typical values range from 250 to 350 ms.

S wave represents the last region of the ventricle muscle to contract and is known as the "late ventricular depolarization."

T wave represents the time for ventricular repolarization, which lasts longer than ventricular depolarization.

QT interval is the combined time for ventricular depolarization and repolarization. Typical values range from 200 to 400 ms.

RR interval is the time difference between two successive QRS complex peak which gives an indication of the instantaneous heart rate at that time. In normal practice, the average heart rate is measured over a number of RR intervals.

Note: Atrial repolarization occurs at the same time as ventricular depolarization and because of the smaller magnitude of the repolarization voltage, its signal level is masked by the relatively larger magnitude of the QRS complex.

The ECG is used to help diagnose problems associated with heart disease and congenital abnormalities in the heart. Irregular heartbeats are known as arrhythmias and can cause the heart to stop beating. Bradycardia arrhythmia is the condition where the heart beats slowly (less than 60 bpm) and tachycardia arrhythmia is when the heart beats quickly (greater than 100 bpm). A dysfunction of the SA node, known as sick sinus syndrome, is one of the most common causes of persistent bradycardia and is caused by a dysfunction of the SA node that can also result in alternate bradycardia and tachycardia rhythms being detected on an ECG.

Heart block occurs when there is a dysfunction of the heart's inherent impulse conduction path from the SA node through the VA node to the bundle of His and Purkinje fibers and is characterized according to the degree of the AV block. First-degree AV conduction block is seen on an ECG waveform as a lengthening of the PR interval to greater than 200 ms, resulting in an abnormal delay between the atria contracting and the ventricles contracting. Second-degree AV block is characterized by two types, Mobitz Type I AV block (also known as Wenckebach's block) where there is a regular atrial rhythm and a progressive lengthening of the PR interval on each heartbeat until eventually the AV node fails to conduct resulting in no ventricular contraction and the QRS effectively "drops out" of the ECG. However, on the next cycle, the PR interval resets to its original length and the cycle repeats. With Mobitz Type II AV block, there is no progressive shortening of the PR interval, but the PR interval is fixed prior to and following the nonconducted beat. Third-degree AV block is known as a complete heart block and is defined when there is no impulse conduction from the SA node to the ventricles. However, the pacemaker cells in the His–Purkinje conduction system will generate an escape ventricular rhythm which will be much slower than the atrial rate.

Pacemakers and implantable cardioverter defibrillators (ICD) are primarily used to maintain or restore a regular heartbeat rhythm which in adults is considered to be between 60 and 80 bpm. Electrical impulses applied to the heart muscles via an implanted electrode can effectively pace the heart in order to restore a more regular heartbeat. However, the generation of two consecutive action potentials in a myocardial cell is limited by the cardiac refractory period (CRP) which is composed of the absolute (also known as the effective) and relative refractory periods. During the absolute refractory period, the cell cannot be depolarized again even in the presence of a strong stimulus. Consequently, no other action potential can be generated. However, during the relative refractory period under certain conditions, a strong stimulus can bring about cell depolarization, resulting in the generation of another action potential, which can lead to cardiac arrhythmia. The CRP has a typical duration of between 200 and 300 ms.

Pacing is used to reset the heart cells back to their refractory state by delivering a series of low-energy cardioversion pulses which are synchronized with the R wave. An initial stimulus will revert the cells in the locality of the stimulation electrode back to their refractory state. Repeated stimulus pulses will then continue to convert other cells further away from the stimulus site into the refractory state, ultimately converting the entire heart to the refractory state. Considering a refractory period of 150−250 ms, the stimulus frequency will be of the order of 5 Hz which can be reduced after reaching the refractory state; this enables the heart to begin beating normally again. In order to achieve a faster recovery time, a normal pacemaker rhythm of typically 70 bpm can be delivered until normal rhythm is achieved.

6.2 **TYPES OF PACEMAKERS**

Pacemakers can be broadly categorized into one of three types; single chamber, dual chamber or biventricular. A single chamber pacemaker with the lead positioned in the right atrium is used when there is an SA node dysfunction but no AV dysfunction, as in the case of sick sinus syndrome. The atrial lead monitors intrinsic atrial activity and delivers an electrical impulse if the SA node beats too slowly. A single chamber pacemaker with the lead placed in the right ventricle is used when the atrial activity does not need to be monitored but the rate of the ventricular activity needs to be managed, as in the case of chronic atrial fibrillation where the atrial activity cannot be controlled by pacing as atrial fibrillation is a rapid rhythm that overrides sinus rate. However, this rhythm can be associated with slow ventricular rates. By placing a pacing lead in the right ventricle, the ventricular rate can be monitored and the pacemaker will initiate a paced beat when the ventricular rate beats too slowly.

Dual chamber pacemakers have one lead inserted in the right atrium and a lead inserted into the right ventricle and are typically used when there is a significant AV block. The dual chamber pacemaker will monitor intrinsic activity in the atria and ventricles and deliver a pacing impulse to either or both chambers as necessary.

Biventricular pacemakers have a third lead typically positioned in a cardiac vein that lines the wall of the left ventricle and are used when the left and right ventricles are not contracting simultaneously. Biventricular pacing is also known as cardio resynchronization therapy (CRT) and uses an electrical stimulus to provide a more balanced and normal contraction of the ventricles in the case of patients with heart failure.

In all modern pacemakers, the implanted electrode is also used to continuously monitor the electrical activity of the heart and to deliver an artificial pacemaker stimulus on demand to pace the heart. The pacemaker implantable device is surgically inserted into a subcutaneous pouch located in the upper chest region, normally below the collarbone. The electrodes are normally inserted into the heart through the cephalic or subclavian veins.

6.3   REVISED NASPE/BPEG GENERIC CODE FOR ANTIBRADYCARDIA PACING

The North American Society of Pacing and Electrophysiology (NASPE) and British Pacing and Electrophysiology Group (BPEG) have collectively adopted the North American and British Generic (NBG) Pacemaker mode table for antibradycardia pacing which defines the modes of operation for most manufactured pacemakers with multiple functionality. Table 6.1 provides the revised NASPE/BPEG Generic (NBG) Pacemaker codes for antibradycardia pacing.

Table 6.1 NBG Pacemaker Codes for Antibradycardia Pacing (Bernstein et al., 2002)

Position	I	II	III	IV	V
Category	Chamber(s) paced	Chamber(s) sensed	Response to sensing	Rate modulation	Multisite pacing
	O = None	**O** = None	**O** = None	**O** = None	**O** = None
	A = Atrium	**A** = Atrium	**T** = Triggered	**R** = Rate Modulation	**A** = Atrium
	V = Ventricle	**V** = Ventricle	**I** = Inhibited		**V** = Ventricle
	D = Dual (A + V)	**D** = Dual (A + V)	**D** = Dual (T + I)		**D** = Dual (A + V)
Manufacturer's designation only	**S** = single (A or V)	**S** = single (A or V)			

The code consists of five letters, the first of which is the position letter which represents a chamber-specific category for pacing, sensing, response to sensing, rate modulation, and multisite pacing. Positions I and II indicate which chambers are to be paced and sensed. Position III relates to whether a pacing stimulus is either triggered or inhibited in response to a sensed cardiac depolarization. Position IV is used to determine whether the pacing rate can be automatically modulated (adjusted) by the pacemaker. The pacemaker can be configured for adaptive pacing to detect physiological parameters related to respiratory, body movement, temperature, oxygen saturation, and pH in the blood, in response to exercise such that the heart rate can be increased in order to cope with the extra physiological demands. Position V indicates whether multisite pacing is available in one or more of the chambers.

There are essentially four pacing modes: asynchronous, single chamber synchronous (demand), dual chamber sequential, and rate-responsive.

Examples of pacemaker settings:

> AOO—Atrial asynchronous pacemaker.
> Pace in atria, no chamber sensing. Atrial pacing at a fixed rate that is independent of the inherent heart rate.
> VOO—Ventricle asynchronous pacemaker.
> Pace in ventricle, no chamber sensing. Ventricular pacing at a fixed rate that is independent of the inherent heart rate.
> AAI—Atrial single chamber synchronous pacemaker.
> Pace in atrium and sense in atrium. Atrium pacing inhibited in response to atrium activity.
> VVI—Ventricular single chamber synchronous pacemaker.
> Pace in ventricle and sense in ventricle. Ventricular pacing inhibited in response to sensing ventricular activity.
> VDD—Ventricular dual chamber synchronous pacemaker.
> Pace in ventricle and sense in both atrial and ventricle. Ventricular pacing is triggered or inhibited in response to sensed atrial and ventricular activity. Sensed atrial activity will trigger ventricle pacing. However, a sensed ventricular activity will inhibit ventricular pacing.
> DVI—Dual chamber ventricle pacing sequential pacemaker.
> Pace in atrium and ventricle and sense in ventricle. Sensed atrial activity triggers a sequence of atrial stimulation, adjustable PR interval, and ventricular stimulation. Ventricle pacing is inhibited in response to ventricular activity.
> DDD—Dual chamber pacemaker.
> Pace in both atrium and ventricle, sense in both atrium and ventricle. Atrium and ventricle pacing inhibited in response to sensing atrial or ventricular activity. Atrium pacing inhibited in response to atrium activity. Trigger ventricular pacing if atria activity is sensed unless ventricular activity is sensed.

VVIRV—Biventricular rate-responsive pacemaker.
Pace in ventricle, sense in ventricle. Ventricular pacing inhibited in response to sensing ventricular activity. Automatic rate modulation is enabled and employs multisite ventricular pacing.

Other rate-responsive pacemakers include AAIR, DDIR, and DDDR.

Pacemakers are now programmable to allow a multitude of parameters to be changed, such as pacing rate, PR interval, mode of pacing, hysteresis, refractory period, stimulus amplitude, stimulus pulse width, and atrial tracking rate to be modified using an inductive coupling or wireless transmission link. Some pacemakers also incorporate a magnetic switch and are programmed to respond when a clinical magnet is placed over the pacemaker. The response is normally to reset the pacemaker to its default asynchronous pacing mode or to initiate a built-in diagnostics mode, after which pacing is resumed.

6.4 IMPLANTABLE CARDIOVERTER DEFIBRILLATORS

Pacemakers are used to sense and correct for bradycardia arrhythmias (slow rhythms) and impulse conduction blocks by applying a low-level electrical pacing stimulus to one or more of the heart chambers, whereas ICDs are used to detect and correct for tachycardia arrhythmias (fast rhythms) by applying a low-energy, high-rate electrical stimulus or a high-energy shock to the heart muscle. Currently, ICDs incorporate pacemaker functionality together with sophisticated techniques to monitor electrical activity in the heart, to provide stimulated shocks, and also to store and transmit ECG data using wireless communication. ICDs differ from pacemakers in that they sense for the ventricular arrhythmias; ventricular tachycardia (VT) and ventricular fibrillation (VF), both of which are life threatening and associated with sudden death.

VT is characterized by a heart rate above 100 bpm which is generated in the ventricles from a single origin (monomorphic) or from a few specific origins (polymorphic). During VT, the ventricles may depolarize before the atria have pumped blood into the ventricles such that the heart chambers do not fill to their full capacity and subsequently less blood is pumped around the body. Slower VTs (100–150 bpm) are generally tolerated by the patient but this depends on the pumping function of the heart. Faster VTs (150–200 bpm) are poorly tolerated and can result in a more rapid hemodynamic compromise such that the patient's blood pressure can fall, causing the patient to black out.

VF is characterized by spontaneous depolarizing of myocardial cells from many different sites within the ventricles, resulting in fast heart rates approaching 300 bpm and the ventricles "quivering" instead of contracting. This leads to very little oxygenated blood being pumped around the body. Fast VTs can lead to VF, leading to sudden cardiac arrest (SCA) and subsequent cardiac death.

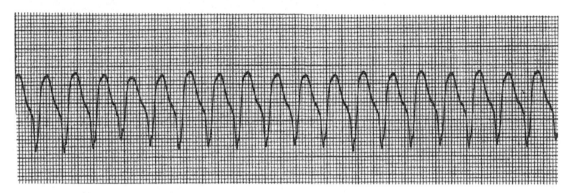

■ **FIGURE 6.3** Ventricular tachycardia.

In order to correct for ventricular arrhythmias, ICDs normally have three standard modes of operation; antitachycardia pacing (ATP) which consists of synchronized low-energy, fast-rate pacing impulses to disrupt the tachycardia circuit, synchronized high-energy defibrillation shock therapy (cardioversion) and unsynchronized high-energy defibrillation shocks. Figure 6.3 shows the ECG for ventricular arrhythmia. If the arrhythmia is fast and regular, a low-level ATP pulse can be delivered to restore the arrhythmia into a normal sinus rhythm. The ICD categorizes VTs into different zones in which different treatment options can be programmed. Faster VTs are less likely to respond to ATP and will require a synchronized shock to terminate the arrhythmias. In the fast VT zone, ATP may be used but more commonly, the ICD enters the cardioverter mode in which a synchronized high-energy shock is delivered to effectively stun the heart and allow normal sinus rhythm to resume.

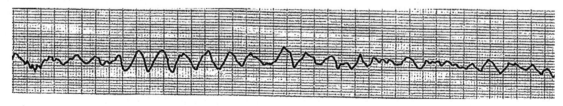

■ **FIGURE 6.4** Ventricular fibrillation.

Figure 6.4 shows the ECG for VF in which the arrhythmia is very irregular and very fast. In this case, the ICD acts as a defibrillator, delivering a high-energy shock to the ventricles in order to restore the natural rhythm of the heart.

6.5 NASPE/BPEG DEFIBRILLATOR CODE

Table 6.2 NBD Code (Bernstein et al., 1993)

Position	I	II	III	IV
Category	Shock chamber	ATP chamber	Tachycardia detection	Antibradycardia pacing chamber
	O = None	**O** = None	**E** = Electrogram	**O** = None
	A = Atrium	**A** = Atrium	**H** = Hemodynamic	**T** = Triggered
	V = Ventricle	**V** = Ventricle		**I** = Inhibited
	D = Dual (A + V)	**D** = Dual (A + V)		**D** = Dual (T + I)

The NASPE/BPEG Defibrillator (NBD) code (Table 6.2) is compatible with the pacemaker NBG code and has been adopted by manufacturers to define the modes of operation for ICD devices. The code consists of four letters, each one representing a specific category for: shock chamber, ATP chamber, means of tachycardia detection and antibradycardia pacing chamber. Positions I, II, and IV indicate which chambers are to be shocked or selected for either ATP or antibradycardia pacing. Position III relates to the devices used for detection of a tachycardia event, an electrogram signal, or one or more hemodynamics-related variables such as blood pressure or transthoracic impedance. It is assumed that all ICDs use electrograms for detection of tachycardia. Selection of hemodynamics is hierarchical in that if selected, then an electrogram is automatically assumed.

Table 6.3 Short Form of the NBD Code (Bernstein et al., 1993)

ICD-S	ICD with shock capability only
ICD-B	ICD with bradycardia with pacing
ICD-T	ICD with tachycardia (and bradycardia) pacing as well as shock

Table 6.3 provides a short form of the NBD code which assumes that a defibrillator with ATP also has the capability of antibradycardia pacing.

6.6 **IMPLANTABLE CARDIOVERTER DESIGN**

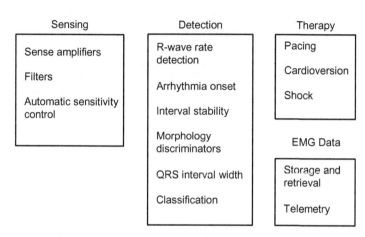

■ **FIGURE 6.5** Elements of a typical ICD system.

A block diagram showing the many elements that comprise a typical ICD is shown in Figure 6.5. Most ICDs now incorporate pacemaker functionality to detect for heart arrhythmias such as bradycardia and tachycardia and to apply the appropriate stimulus therapy to return the heartbeat to a more normal rhythm. As the QRS complex is the most dominant feature in the ECG waveform, it is used to determine the heart rate and to provide a measure of ventricular activity. The heart rate can be measured by detecting the time interval between successive R-R intervals and comparing against known interval times, categorizing the heart rate as too fast, normal, or too slow. Subsequently, the P and T waves of the ECG waveform are normally filtered out together with unwanted biomechanical artifact noises such as muscle movement and any external electromagnetic noise interference (EMI). EMI noise from external radiated sources can interfere with the sensed ventricular signals which can erroneously be sensed as VT or VF. By separating out the QRS complex from the ECG waveform, the QRS features can be extracted for analysis and subsequent arrhythmia classification.

The sensitivity of the sensing amplifier is designed to automatically adapt to sensing the large amplitude QRS complexes in preference to the relatively smaller T waves and also to adjust the amplifier's sensitivity when sensing smaller amplitude-persistent tachycardia arrhythmia. The sensitivity control or tracking is achieved by either incorporating an automatic gain control amplifier whereby the gain increases for small signals or vice versa using a fixed threshold level or using automatic threshold tracking which adjusts the threshold level, depending on the sensed signal level. When the signal level decreases, the threshold level decreases and vice versa. Any excursions above the threshold level are registered as sensed events.

The comparison of the amplitude of sensed arrhythmia events (episodes) to that of a threshold level, gives an indication of the number of cycles or episodes and starts a counter to determine the number of cycles.

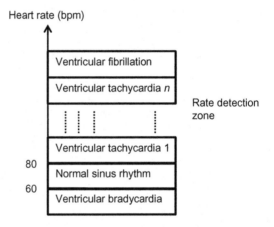

■ **FIGURE 6.6** Heart rate detection zones.

Rate detection is based upon the number of times a measured electrogram (EGM) parameter such as the R-R interval falls within predefined limits within defined rate detection zones, as shown in the example in Figure 6.6. Each zone represents a range of heart rates associated with bradycardia, normal sinus rhythm, degrees of VT and VF. All ranges are defined within boundary limits that can be preprogrammed. When a number of consecutive, rate-detected intervals exceed a predetermined count within a zone, the appropriate pacing or shock therapy is initiated. Associated with a QRS complex is a 200 ms refractory period which is due to the physiological makeup of the heart. Subsequently, no signal detection occurs during this period.

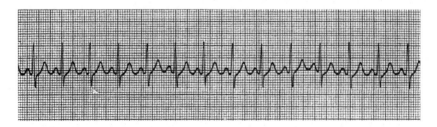

■ **FIGURE 6.7** Sinus tachycardia.

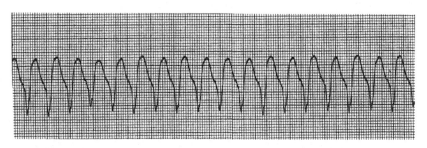

■ **FIGURE 6.8** Ventricular tachycardia.

Tachyarrhythmias can also originate in the atria and are known as supra-ventricular tachyarrhythmias (SVT) which include; atrial fibrillation, sinus tachycardia (ST) and atrial flutter. ST is a normal fast heart rhythm which occurs during exercise and is characterized by a heart rate of greater than 100 bpm which could erroneously be detected as a ventricular arrhythmia requiring therapy. Consequently, ICDs use a number of enhancements and classification techniques to supplement rate detection in order to help discriminate between ST (Figure 6.7) and VT (Figure 6.8). These enhancements focus on the difference in the arrhythmia patterns between ST and VT. Other SVTs are seen when impulses in the atria overdrive the sinus node activity, e.g., atrial fibrillation and atrial flutter. Atrial fibrillation has an irregular cardiac rhythm compared to VT which has a relatively more stable arrhythmia whereas ST has a gradual onset compared to VT. Also, a wider QRS interval is normally associated with VT rather than ST or SVTs. Some ICDs incorporate morphology discriminator algorithms that are used to compare the features of

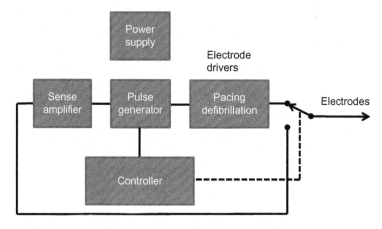

■ **FIGURE 6.9** Block diagram of a typical ICD incorporating a pacemaker.

the sensed QRS complex against a stored baseline normal sinus rhythm template in order to discriminate and classify VT episodes. Dual chamber ICDs enable atrial sensing as well as ventricular sensing to determine ventricular arrhythmia more accurately.

Figure 6.9 shows a general block diagram of a single channel ICD. Combined pacemaker and ICDs normally have two separate sense amplifiers, one designed for bradycardia detection and the other for tachycardia detection. The high-energy density batteries in ICDs are of low voltage types, typically up to 3 V and therefore, a step-up energy conversion circuit such as a DC−DC converter, flyback transformer and voltage doubler/tripler diode network are implemented to achieve a standard output voltage of 750 V or more in order to deliver high-energy pulses (typically 30−40 J) to disrupt VF and return normal sinus rhythm. The electrode driver circuits charge a capacitor or bank of capacitors which when discharged, deliver the high-energy pulses. Consequently, the sensing amplifiers incorporate protection circuits to protect the sensing electrodes from the high-energy voltage pulses.

6.7 MEDTRONIC MICRA TRANSCATHETER PACING SYSTEM

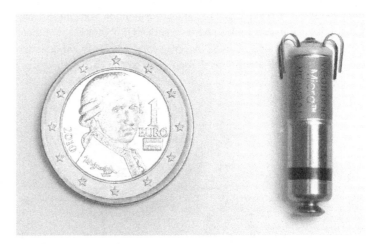

■ **FIGURE 6.10** Micra TPS. *(Copyright © Medtronic. Reprinted with permission.)*

The Micra™ Transcatheter Pacing System (TPS) (Figure 6.10) measures 25.9 mm and is smaller by about one tenth the size of a conventional pacemaker. The TPS is inserted into the heart via a catheter inserted in the femoral vein. One end of the TPS contains tines which are used to attach the TPS to the heart wall and to also allow for the device to be repositioned if needed. The TPS is leadless and self-contained in that stimulation current is delivered by an electrode at the distal end of the TPS compared to a conventional pacemaker where electrode leads are threaded into the heart and attached to a pacemaker that is implanted subcutaneously in the chest region.

6.8 MEDTRONIC VIVA AND EVERA

■ **FIGURE 6.11** Viva® XT CRT-D. *(Copyright © Medtronic. Reprinted with permission.)*

The Viva® XT device (Figure 6.11) from Medtronic provides cardiac resynchronization therapy with defibrillation (CRT-D) incorporating the AdaptivCRT™ algorithm which continuously adapts to individual patient needs by optimizing the CRT pacing to preserve patient's normal heart rhythms and reduce the symptoms associated with heart failure.

■ **FIGURE 6.12** Evera® XT MRI DR ICD. *(Copyright © Medtronic. Reprinted with permission.)*

The Viva XT CRT-D and the Evera® DR XT Implantable Cardioverter Defibrillator (ICD) (Figure 6.12) devices both utilise SmartShock technology that consists of six exclusive algorithms that discriminate between true lethal arrhythmias from other arrhythmic and non-arrhythmic events. Evera® ICDs also incorporate SureScan™ technology which when used with SureScan™ CapSureFix MRI® leads, are considered MR-conditional devices. This allows patients to safely undergo full body MRI scans.

The devices also incorporates Medtronic's proprietary PhysioCurve™ design to reduce skin pressure by providing a more contoured shape to better fit the body.

6.9 **SORIN GROUP KORA 100**

The KORA 100 MRI pacing system from Sorin Group consists of the KORA 100 SR and KORA 100 DR pacemakers (Figure 6.13), which when implanted with the Sorin BEFLEX pacing lead, allows patients to undergo magnetic resonance imaging (MRI) safely. The Automatic MRI Mode detects the MRI scanner's strong magnetic field and automatically switches the device to operate in asynchronous mode. Similarly, the device senses when a patient leaves the magnetic field and switches back to normal mode within 5 min, thus limiting the amount of time that the pacemakers operate in MRI mode. The KORA 100 pacemakers also incorporate proprietary Sleep Apnea Monitoring which automatically screens patients for severe sleep apnea.

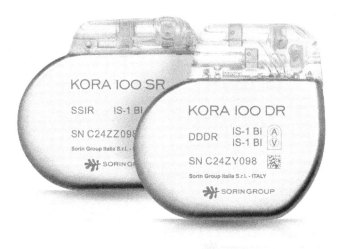

■ **FIGURE 6.13** The KORA 100 SR and 100 DR pacemakers. *(Copyright © Sorin Group. Reprinted with permission.)*

The Sorin Group CRT-D optimization system includes the SonR hemodynamic sensor embedded in the SonRtip atrial pacing lead and the SonR CRT-D device (Paradym RF SonR or Intensia SonR). SonR is a weekly self-adjusting CRT-optimization system providing optimized timing at rest and exercise for improved CRT response (Figure 6.14).

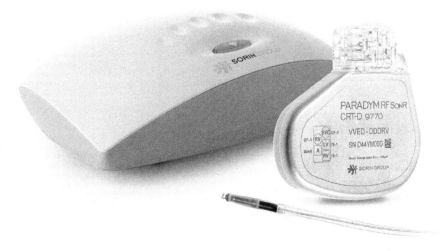

■ **FIGURE 6.14** Smartview, Paradym RF SonR. *(Copyright © Sorin Group. Reprinted with permission.)*

6.10 **BIOTRONIK**

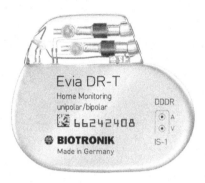

■ **FIGURE 6.15** Evia DR-T pacemaker. *(Copyright © Biotronik. Reprinted with permission.)*

The Evia family of pacemakers from Biotronik consists of single, dual, and CRT devices incorporating proprietary ProMRI technology that allows patients conditional access to MRI examinations. Figure 6.15 shows the Evia DR-T pacemaker. Home Monitoring® technology provides continuous monitoring of clinical parameters and early detection of atrial and ventricular arrhythmia, device and lead problems. Closed Loop Stimulation (CLS) offers physiological rate response and treatment options for patients with MVVS. CLS integrates into the natural cardiovascular loop which initiates pacing earlier and more effectively, especially during periods of emotional or mental stress. CLS provides optimized heart rate response, resulting in improvement or restoration of patients' quality of life.

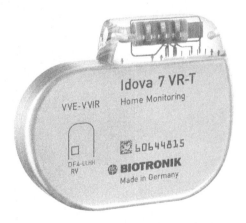

■ **FIGURE 6.16** Idova VR-T pacemaker. *(Copyright © Biotronik. Reprinted with permission.)*

The Idova 7 ICD family consists of single chamber, dual chamber, CRT, and Dx (single chamber with atrial sensing) devices, each capable of delivering high-energy shocks, especially for patients with high defibrillation thresholds. Figure 6.16 shows the Idova VR-T pacemaker. Proprietary ProMRI technology allows patients conditional access to MRI examinations and Home Monitoring® technology provides continuous monitoring of clinical parameters and early detection of atrial arrhythmias. SelectSense® sensing adapts sensing parameters to patients' individual needs via a sophisticated automatic sensitivity control algorithm, and SafeSync® provides wireless data telemetry via an external wand.

6.11 ST JUDE MEDICAL NANOSTIM™

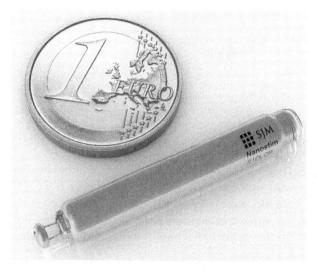

■ **FIGURE 6.17** Nanostim™ leadless pacemaker. *(Copyright © St Jude Medical. Reprinted with permission.)*

The Nanostim™ is a leadless VVIR pacemaker which is inserted into the heart through the femoral vein and implanted into the right ventricle where it is fixed in place with a dual fixation system (Figure 6.17). The Nanostim™ incorporates a steroid-eluting electrode at one end which is used to stimulate and pace the heart. The Nanostim™ incorporates an internal battery with a life expectancy of 9 years at 100% pacing and 13 or more years at 50% pacing. The Nanostim™ has also been designed so that it can easily be repositioned or retrieved in case of battery replacement.

At the time of writing, the Nanostim™ is not commercially available.

6.12 **ST JUDE UNIFY QUADRA**™ **AND ACCENT**™

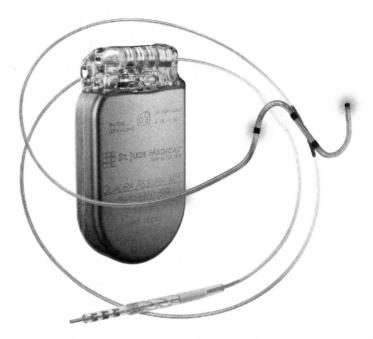

■ **FIGURE 6.18** Unify Quadra™ CRT-D system. *(Copyright © St Jude Medical. Reprinted with permission.)*

The Unify Quadra™ CRT-D (Figure 6.18) system uses quadripolar pacing technology and includes the Quartet Left Ventricular (LV) lead which has four electrodes and the VectSelect™ Quartet LV Pulse Configuration to provide 10 pacing configurations. Four electrodes can cover a greater area such that their precise placement is less critical and stimulation currents can be adjusted by software control.

■ **FIGURE 6.19** Accent™ MRI pacemaker. *(Copyright © St Jude Medical. Reprinted with permission.)*

The Accent MRI™ pacemaker (Figure 6.19) and Tendril MRI™ pacing lead enable patients to safely undergo a full-body MRI scan. The external Activator™ held unit can be used to enable and disable preprogrammed MRI settings. The Accent MRI™ pacemaker also incorporates InvisiLink™ wireless telemetry for remote patient monitoring. Other features include QuickOpt™ for timing cycle optimization and SenseAbility™ technology to deliver optimal therapy for patients at implant and throughout their lives.

6.13 **BOSTON SCIENTIFIC INGENIO™ AND INCEPTA™**

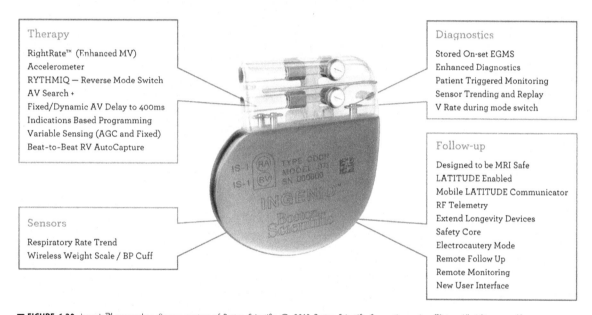

Therapy

RightRate™ (Enhanced MV)
Accelerometer
RYTHMIQ — Reverse Mode Switch
AV Search +
Fixed/Dynamic AV Delay to 400ms
Indications Based Programming
Variable Sensing (AGC and Fixed)
Beat-to-Beat RV AutoCapture

Sensors

Respiratory Rate Trend
Wireless Weight Scale / BP Cuff

Diagnostics

Stored On-set EGMS
Enhanced Diagnostics
Patient Triggered Monitoring
Sensor Trending and Replay
V Rate during mode switch

Follow-up

Designed to be MRI Safe
LATITUDE Enabled
Mobile LATITUDE Communicator
RF Telemetry
Extend Longevity Devices
Safety Core
Electrocautery Mode
Remote Follow Up
Remote Monitoring
New User Interface

■ **FIGURE 6.20** Ingenio™ pacemaker. *(Image courtesy of Boston Scientific. © 2013 Boston Scientific Corporation or its affiliates. All rights reserved.)*

Figure 6.20 shows the Ingenio™ pacemaker from Boston Scientific which uses their proprietary RightRate™ minute ventilation sensor to restore chronotropic competence and RYTHMIQ™, an algorithm that minimizes unnecessary RV pacing without clinically significant pauses. INGENIO™ allows the heart to take the lead, intervening only when appropriate.

Figure 6.21 shows the Boston Scientific INCEPTA™ ICD advanced system solution offering patient comorbidities monitoring (ApneaScan™, HF monitoring in DR, LATITUDE™ Patient Management with sensors). The unique ApneaScan™ feature is designed to identify patients at risk of sleep apnea, providing a broader perspective on the clinical status of patient. These small, thin, high-energy devices with an increased and predictable longevity

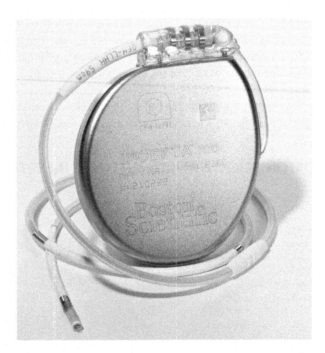

■ **FIGURE 6.21** INCEPTA™ ICD. *(Image courtesy of Boston Scientific. © 2013 Boston Scientific Corporation or its affiliates. All rights reserved.)*

up to 8 years are designed for better patient acceptance, comfort and increased quality of life. The new Acushock arrhythmia discrimination system features high specificity and therapy algorithm: RHYTHM ID with RHYTHM MATCH™ is designed for minimization of inappropriate therapy, allowing flexible customization of arrhythmia detection. Avoiding unnecessary right ventricular pacing using RYTHMIQ™, and delivering the right therapy at the right time, INCEPTA™ ICD ensures patient-tailored treatment. Featuring HF Perspectiv, including Respiratory Rate Trend and a complete HRV diagnostic monitoring in DR model, in combination with LATITUDE™ Patient Management System, INCEPTA™ ICD provides an advanced solution for patient management.

6.14 **BOSTON SCIENTIFIC SUBCUTANEOUS ICD**

The subcutaneous ICD (S-ICD) System (Figure 6.22) from Boston Scientific provides a new therapeutic solution for patients at risk of sudden cardiac arrest while leaving the heart and vasculature untouched. Similar to transvenous ICDs, the S-ICD™ System utilizes a pulse generator capable of

■ **FIGURE 6.22** S-ICD™ System. *(Image courtesy of Boston Scientific. © 2013 Boston Scientific Corporation or its affiliates. All rights reserved.)*

delivering life-saving high-energy shock therapy for VT and VF. Unlike transvenous ICDs, the S-ICD™ System is implanted in the lateral thoracic region of the body and utilizes a subcutaneous electrode instead of transvenous leads to both sense and deliver therapy.

BIBLIOGRAPHY

Bernstein, A.D., et al., 1993. North American Society of Pacing and Electrophysiology policy statement. NASPE/BPEG defibrillator code. Pacing Clin. Electrophysiol. 16 (9X), 1776–1780.

Bernstein, A.D., et al., 2002. The revised NASPE/BPEG generic code for antibradycardia, adaptive-rate, and multisite pacing. North American Society of Pacing and Electrophysiology/British Pacing and Electrophysiology Group. Pacing Clin. Electrophysiol. 25 (2), 260–264.

Chapter

Bladder Implants

7.1 INTRODUCTION

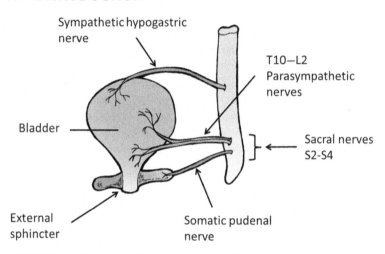

■ **FIGURE 7.1** Simplified diagram of neural innervations to the bladder and external sphincters.

The neural pathways innervating bladder function originate from the S2 to S4 sacral roots of the spinal cord between the T12 to L1 vertebral bodies. These sacral segments form the pelvic nerves, providing parasympathetic, sympathetic, and somatic pathways, innervating the bladder as shown in Figure 7.1. The external sphincter is made up of the musculature of the pelvic floor and is innervated by the somatic pudendal nerve. Stimulation of the parasympathetic nerves results in a contraction of the bladder wall (detrusor muscle) and relaxation of the internal sphincter as the urine is pushed out of the bladder and through the external sphincter. Stimulation of the sympathetic (hypogastric) nerves results in an inhibitory effect, causing relaxation of the bladder wall and an excitatory input to the external sphincter. Stimulation of the pudendal nerve will cause the external sphincter to contract. The external sphincter can also be closed by voluntary effort but relaxes by reflex action as soon as urine passes through the internal sphincter.

Implantable Electronic Medical Devices. DOI: http://dx.doi.org/10.1016/B978-0-12-416556-4.00007-3

When the bladder fills up with urine, stretch receptors in the bladder wall send afferent neural signals to the spinal cord and to the higher centers via the pelvic nerves. As the bladder becomes more distended, reflex contractions become more frequent, stronger, and prolonged. The contractions are accompanied by an increase in the neural activity of the pudendal nerve which causes the external sphincter to contract, promoting continence. When there is a desire to micturate (also known as voiding), parasympathetic activity decreases, causing contraction of the bladder wall and relaxation of the external sphincter. The external sphincter, under control from higher centers, is reflex inhibited via the pudendal nerve. The contraction of the detrusor muscle then pushes urine out of the bladder and through the relaxed internal and external sphincters, causing micturition. When the bladder is empty, the bladder wall relaxes and the sphincters close. The whole process of the bladder filling and emptying, without control from the higher centers, is also known as a sacral reflex arc. If there is no desire to micturate, neural signals transmitted from the higher centers cause parasympathetic activity to decrease and sympathetic activity to increase, causing the bladder wall to relax and the external sphincter to contract. This results in an increase in bladder pressure, providing voluntary continence overriding the natural sacral reflex arc for micturition or voiding.

There are many bladder disorders which can benefit from a biomedical implant, including detrusor hyperreflexia, detrusor areflexia, urinary retention and overactive bladder.

7.2 DETRUSOR HYPERREFLEXIA

Detrusor hyperreflexia is a condition in which the bladder still functions but there is loss of voluntary control normally associated with neurological spinal cord lesions above the sacral cord segments such that the sacral reflex is intact and functioning but is not under higher center control. The bladder will still fill up with urine but periodic involuntary contractions will cause the bladder to spontaneously empty. The micturition will not normally be complete, leaving a small residual volume of urine in the bladder. However, if the relaxations of the external sphincter are coordinated with the bladder wall contractions, bladder evacuation may be adequate, providing continence. If there is uncoordinated activity of the external sphincter with the bladder, known as detrusor—sphincter dyssynergia, pressure inside the bladder will rise due to uncoordinated contractions of the external sphincter causing the bladder to empty frequently, leading to active incontinence. This results in a residual volume of urine which can be quite large and can often lead to increased risk of infections and possible kidney damage due to an increase in back pressure effects.

7.3 **DETRUSOR AREFLEXIA**

This disorder occurs when there is a loss of reflex bladder control leading to the bladder wall becoming flaccid and increasing in capacity with no sensation of increased bladder pressure. Subsequently the bladder overflows, resulting in continual dribbling. This condition normally occurs when there is a spinal cord lesion at the sacral segment level, damage to the cauda equina or damage to the pelvic nerves.

Voluntary emptying of the bladder is possible by manual compression of the anterior abdominal wall, resulting in the external sphincter relaxing but a small residual volume of urine will still be left in the bladder. Alternatively, a catheter inserted into the urethra provides an efficient method of emptying the excess urine.

7.4 **OVERACTIVE BLADDER SYNDROME AND URINE RETENTION**

Overactive bladder syndrome (OAB) is a condition that causes the bladder to contract involuntarily even when not full and gives a sense of urgency to urinate. Urge-incontinence is a condition that causes a leakage of urine as the bladder contracts before or at the same time there is the urge to urinate. These OAB contractions and sudden urges to urinate, lead to frequent visits to the toilet. With OAB contractions, the bladder will hold less urine with more frequent urges to urinate.

The International Continence Society defines OAB syndrome as the urgency (sudden compelling and unstoppable desire) to urinate with or without urge-incontinence. The urgency-frequency refers to the number of times a person urinates during the day (more than seven times) and defines nocturia as the number of times a person has to wake up (more than once) to urinate during the night.

Urinary retention is the inability to urinate in order to empty or partially empty the bladder. This condition can arise from an obstruction such as kidney stones or a nonobstructive cause such as a stroke or neural trauma leading to a loss of neural innervation to the bladder or a weakness of the musculature of the bladder wall.

7.5 **SACRAL ANTERIOR ROOT STIMULATION**

The pelvic nerves are not easily accessible for placing stimulating electrodes. However, the parasympathetic and somatic pathways originate from the same sacral roots. The parasympathetic nerves innervating the bladder wall are smaller in diameter to the larger somatic nerve fibers innervating the external sphincter with no distinguishable overlap in fiber size diameter. Stimulation of the sacral roots will therefore evoke responses in both the parasympathetic and somatic nerve fibers. However, conventional stimulating electrodes exhibit an unnatural inverse recruitment order of nerve fibers based on their fiber size diameter. The larger diameter somatic fibers will be stimulated prior to the smaller diameter parasympathetic fibers, resulting in the contraction of the external sphincter before the bladder wall has had time to contract. Consequently, the natural physiological coordination of bladder contraction and external sphincter relaxation is effectively reversed, leading to urine retention, kidney back pressure, and infection, which subsequently limits the effectiveness of using conventional stimulation techniques.

7.6 FINETECH-BRINDLEY SACRAL ANTERIOR ROOT STIMULATORS, FINETECH MEDICAL LTD.

The Sacral Anterior Root Stimulators (SARS) from Finetech Medical Ltd. takes advantage of the difference in the speeds of muscle contraction between the striated muscle of the external sphincter and the smooth muscle of the bladder wall (Brindley, 1977). As the external sphincter contracts and relaxes much faster compared to the smooth muscle of the bladder wall, the time between stimulation pulses is set so that the external sphincter will contract quickly during each pulse duration and relax quickly between pulses, whereas the smooth bladder wall will slowly contract during each pulse duration and slowly relax during the interpulse interval (Figure 7.2).

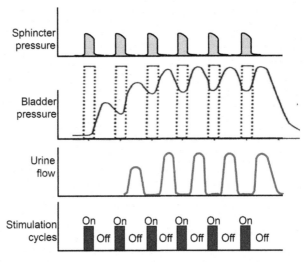

■ **FIGURE 7.2** SARS stimulation strategy to affect micturition. *(Copyright © Finetech Medical Ltd. Reprinted with permission.)*

If the interpulse interval is not too long, the bladder wall will continue to contract such that intermittent micturition (urine flow) will then occur during the interpulse intervals when the stimulation is turned off. During system implantation, the dorsal roots of the sacral segments are normally cut in order to reduce the effects of any unwanted afferent impulses causing sacral reflex actions and possible sensation of pain.

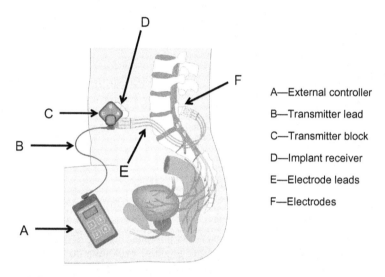

A—External controller

B—Transmitter lead

C—Transmitter block

D—Implant receiver

E—Electrode leads

F—Electrodes

■ **FIGURE 7.3** Finetech-Brindley SARS. *(Copyright © Finetech Medical Ltd. Reprinted with permission.)*

The Finetech-Brindley SARS shown in Figure 7.3 consists of an external digital controller connected to a transmitter block which delivers power and control signals to the subdermally implanted receiver in the abdomen. Cables run from the receiver to the electrodes implanted on the sacral roots.

Figure 7.4 shows the external controller of the Finetech-Brindley stimulator, the transmitter block and sacral root electrodes. The sacral root electrodes consists of a "book type" electrode configuration which for an intrathecal implant, clasps over the three sacral roots, S2 to S4 and is connected to the implanted receiver in the abdomen.

(A)   (B)

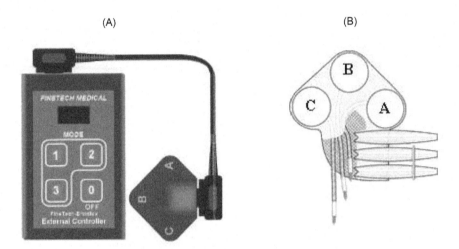

■ **FIGURE 7.4** Finetech-Brindley SARS: (A) external controller and transmitter block and (B) sacral root electrodes. *(Copyright © Finetech Medical Ltd. Reprinted with permission.)*

7.7 MEDTRONIC INTERSTIM® THERAPY

Sacral Neuromodulation with the InterStim® System from Medtronic is used for the treatment of an OAB and urinary retention and may help alleviate urge-incontinence, urinary retention and urgency-frequency by delivering a mild electrical stimulus in close proximity to the sacral nerves (Figure 7.5). Neuromodulation of the sacral nerves has been shown to generate action potentials that modulate the processing of bladder afferent signals in the central nervous system which in turn modulates the abnormal afferent (sensory) activity and thereby influences the neural communication between the bladder and the brain. Consequently, Sacral Neuromodulation is thought to help normalize abnormal sacral nerve activity and to influence the higher centers in the brain responsible for bladder control in order to restore a more normal bladder function.

The InterStim® System from Medtronic consists of a neurostimulator (Figure 7.6) implanted subdermally in the region of the upper buttock and a tined sacral nerve electrode array, or lead, placed near the S3 sacral root in the sacral foramen. There are two generations of InterStim® neurostimulators which both use an external patient controller to control the level of stimulation. InterStim I incorporates a 3.7 V lithium–thionyl chloride cell rated at 2.7 Ah and has a 10.5 V maximum output. InterStim® II incorporates a 3.2 V Lithium/Silver Vanadium Oxide hybrid battery rated at 1.3 Ah

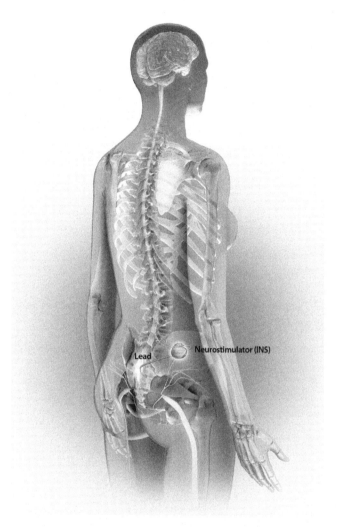

■ FIGURE 7.5 Sacral Neuromodulation using the InterStim system. *(Copyright © Medtronic. Reprinted with permission.)*

and has an 8.5 V maximum output. Both neurostimulators can deliver biphasic stimulation pulses with a pulse width between 60 and 450 μs in steps of 30 μs with a pulse frequency of between 2.1 and 130 Hz. The stimulation pathway can be configured as bipolar or unipolar, in which case the neurostimulator acts as the return electrode.

■ **FIGURE 7.6** InterStim implantable neurostimulator. *(Copyright © Medtronic. Photo courtesy of Medtronic. Reprinted with permission.)*

Figure 7.7 shows the 3889 model tined lead which consists of four equidistantly spaced cylindrical electrode contacts.

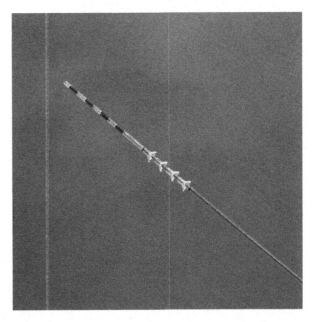

■ **FIGURE 7.7** Tined lead (Model 3889). *(Copyright © Medtronic. Reprinted with permission.)*

BIBLIOGRAPHY

Brindley, G.S., 1977. An implant to empty the bladder or close the urethra. J. Neurol. Neurosurg. Psychiatry 40, 358–369.

Chapter **8**

Electrical Stimulation Therapy
for Pain Relief and Management

8.1 OCCIPITAL NERVE STIMULATION

Occipital nerve stimulation (ONS) uses electrical pulses for the relief of chronic migraines, the cause of which is not clearly understood but is thought to be neurovascular in nature and brought about by external physiological or environmental triggers. Headaches and migraines are usually defined as chronic if they last for more than 15 days per month for more than 3 months. Primary headaches are defined as having no obvious underlying pathological problems whereas secondary headaches exhibit underlying pathological problems. The occipital region of the brain lies at the posterior toward the back of the skull and is innervated by the greater and lesser occipital nerves, which have their origins in the cervical roots, C2 and C3. The mechanisms by which ONS can relieve symptoms of chronic headache are not completely understood but electrical stimulation therapy of the greater occipital nerve has been shown to provide some relief for those suffering with chronic migraines.

8.2 ST JUDE MEDICAL IMPLANTABLE PULSE GENERATORS FOR ONS OF THE OCCIPITAL NERVES

The Implantable Pulse Generators (IPGs) from St Jude Medical provide a range of electrical therapy treatments for intractable which is chronic migraine from simple pain management (Genesis®) to sustained pain therapy stimulation using nonrechargeable (Eon *C*™) and rechargeable (Eon® and Eon® Mini) devices. Figure 8.1 shows the Eon® Mini™, the smallest IPG in the range.

Implantable Electronic Medical Devices. DOI: http://dx.doi.org/10.1016/B978-0-12-416556-4.00008-5

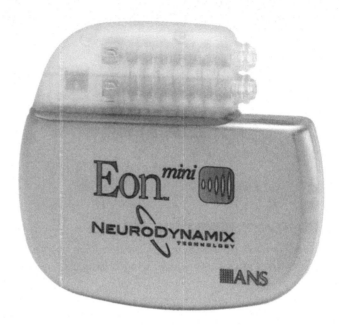

■ **FIGURE 8.1** Eon® Mini™. *(Copyright © St Jude Medical. Printed with permission.)*

All of the IPGs monitor therapy requirements and deliver efficient power management using proprietary NeuroDynamic™ technology. The IPGs incorporate constant current source outputs such that any change detected in the electrode-neural impedance results in an adjustment of the output voltage in order to maintain a constant applied electrical stimulation field. The percutaneous leads consist of either four or eight electrode contacts which are placed in contact with the occipital nerve. An external Rapid Programmer™ system incorporates MultiSteering™ technology to help locate the desired stimulation area and effectively steer the electrode current to that area to optimize the delivered stimulation therapy. The Eon® Mini™ product does not have FDA approval for use in the United States.

8.3 **BOSTON SCIENTIFIC PRECISION SPECTRA™ SCS SYSTEM**

Figure 8.2 shows the Precision Spectra™ SCS rechargeable system which is used to provide pain relief for patients who suffer with chronic pain. The system consists of an IPG which is implanted in the abdomen, lower buttock or below the clavicle and an implantable electrode array

■ **FIGURE 8.2** Precision Spectra™ SCS. *(Copyright © Boston Scientific. Reprinted with permission.)*

implanted alongside the spinal cord, consisting of 32 contacts arranged as either four 8-contacts or two 16-contact leads, allowing for greater coverage area. Electrical impulses sent to the electrodes are perceived as a smooth tingling sensation known as paresthesia so that the feeling of pain may be reduced.

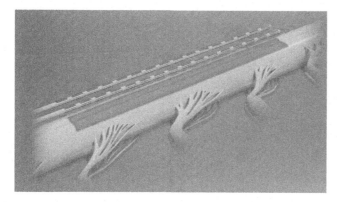

■ **FIGURE 8.3** Placement of epidural electrodes. *(Copyright © Boston Scientific. Reprinted with permission.)*

The electrode array is inserted using an epidural needle so that the electrode contacts cover the dorsal columns of the spinal cord where stimulation is required (Figure 8.3). The IPG incorporates Multiple Independent Current Control (MICC) technology whereby each electrode contact has its own independent power source so that electric fields can be set up between electrodes. The IPG is current controlled and provides an automatic adjustment of stimulation voltage at each electrode in order to provide a constant current when the neural−impedance interface changes due to scar tissue formation after implantation.

The external remote control is used to program the IPG via an inductively coupled RF telemetry link to provide up to four stimulation areas per program with up to 16 possible stimulation programs, allowing patients to target and control their own pain management therapy for different pain problems and different postures at different times of the day. The system is programmed by proprietary Illumina 3D™ software which takes into account the complex three-dimensional environment in which the leads exist. The algorithm incorporates three-dimensional lead location and creates a customized stimulation field designed to improve pain targeting. LeadSync™ Technology is able to detect relative lead location and adjusts for lead offset by synchronizing contacts on parallel leads. It has been designed to account for relative lead movement during the trial, after the permanent implant, or during the procedure.

The IPG incorporates an internal Zero-Volt™ rechargeable battery which does not suffer from battery failure when completely discharged and can last up to 5 years, depending on parameters and usage. The battery is externally charged by placing the battery charger over the site of the IPG.

Chapter

9

Electrical Stimulation Therapy for Parkinson's Disease and Dystonia

9.1 INTRODUCTION

Common neurological disorders of the brain that affect motor function and coordination of body movement include Parkinson's disease, dystonia and a movement disorder known as essential tremor and dystonia. Parkinson's disease is a progressive neurological disorder where loss of nerve cells in the substantia nigra region of the brain results in reduced levels of dopamine which enables cells involved in the control and regulation of movement to communicate. The symptoms of Parkinson's include: muscle tremor leading to involuntary shaking of the head, limbs or body, muscle rigidity experienced as muscle stiffness in the limbs or joints, and slow muscle movements known as bradykinesia, which can lead to akinesia, not being able to move at all. The cause of Parkinson's disease is not widely understood and there is no known cure, although medication, therapy, and in some cases surgery, can bring some relief to the symptoms. The cause of dystonia is unknown but is linked to the basal ganglia region of the brain, the area responsible for movement. Dystonia can be classified as primary dystonia, when there is no obvious degeneration or structural change to the brain, or secondary dystonia, which results from brain degeneration or damage caused by trauma, stroke, tumor, infection, or drug-related issues. Symptoms of dystonia include involuntary sustained muscle contractions and muscle spasms leading to painful body contortions. There is no known cure for dystonia, although drug therapy, muscle relaxants, and nerve toxins such as *Botulinum* injected into the muscle can bring some relief to the symptoms.

Implantable Electronic Medical Devices. DOI: http://dx.doi.org/10.1016/B978-0-12-416556-4.00009-7

Essential tremor is another movement disorder in which the cause is unknown and which displays symptoms of fast, rhythmic tremors of the limbs, head and trunk similar to Parkinson's disease. However, the tremors become more pronounced with movement of the affected parts of the body unlike Parkinson's disease in which the tremors occur during periods of rest when there is no movement.

Functional electrical stimulation provides an alternative therapy in the treatment of Parkinson's disease and dystonia, although it is normally used in addition to medication. By delivering an electrical stimulus to parts of the brain responsible for motor function, some control of muscle movement is possible by blocking neural signals which cause symptoms associated with Parkinson's disease and dystonia. Areas of the brain targeted for electrical stimulation include the subthalamic nucleus or the globus pallidus interna. Electrodes are placed on both sides of the brain for bilateral stimulation of these areas. In the case of tremor, the ventral intermediate nucleus of the thalamus is targeted for unilateral stimulation. The electrode stimulation sites are normally mapped and localized using image-scanning techniques such as magnetic resonance imaging and computerised tomography in order that stereotactic surgery can be performed to implant the electrodes.

9.2 VERCISE™ DEEP BRAIN STIMULATOR, BOSTON SCIENTIFIC

The Vercise™ Deep Brain Stimulator (DBS) is used for the treatment of advanced Parkinson's disease and primary and secondary dystonia by providing electrical stimulation therapy to the subthalamic nucleus or the globus pallidus interna regions of the brain. The electrical impulses are delivered using current source steering between electrodes using Multiple Independent Current Control (MICC) where each electrode has its own independent current source. Subsequently, it is possible to control the depth of the current field into the brain tissue between electrodes, and also to reduce unwanted stimulation of adjacent neural tissue. Each current source electrode can be programmed to operate at a different frequency for possible finer current control to target specific areas in the brain.

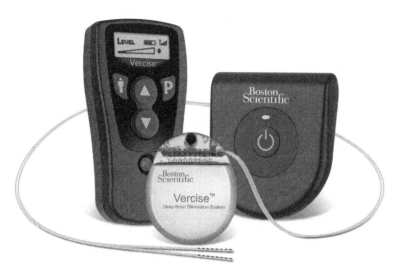

■ **FIGURE 9.1** Vercise™ DBS. *(Copyright Boston Scientific ©. Reprinted with permission.)*

The Vercise™ DBS system consists of an implantable stimulator, leads, remote control and a wireless battery charging unit (Figure 9.1). The rechargeable battery has an expected battery life of 25 years and incorporates proprietary Zero Volt™ technology such that the battery can be completely discharged without subsequent battery failure or damage. The implantable stimulator is also contoured to fit more comfortably to the shape of the cranium. The stimulation therapy consists of either monophasic or biphasic current pulses with a programmable pulse width between 10 and 450 μs, maximum amplitude of 20 mA and a stimulus frequency between 2 and 255 Hz. The wireless remote control unit allows for the stimulation parameters to be adjusted for optimal performance. The electrode array consists of two leads, each with eight ring electrode contacts. The electrodes cover a span of 15.5 mm with a 1.5 mm contact length and 0.5 mm spacing between electrode contacts. The electrode lead has a multilumen helix structure and has been designed to be durable to prevent cables from shorting and for accurate placement.

9.3 **MEDTRONIC ACTIVA PC+S DBS**

The Activa® PC+S system is an investigational DBS system currently being used for research purposes for the treatment of Parkinson's disease,

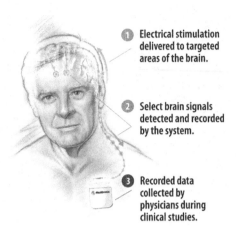

1. Electrical stimulation delivered to targeted areas of the brain.

2. Select brain signals detected and recorded by the system.

3. Recorded data collected by physicians during clinical studies.

■ **FIGURE 9.2** Activa® PC+S DBS system. *(Copyright © Medtronic. Reprinted with permission.)*

dystonia, essential tremor, and treatment-resistant obsessive–compulsive disorder (OCD). The system consists of an implantable neurostimulator, two leads each with four electrodes, and two implantable extension leads. Electrodes are inserted into the brain and connected behind the ear under the scalp to lead extensions that travel under the skin and down the neck to the stimulator, which is implanted subcutaneously in the upper chest area near the clavicle (Figure 9.2). The leads are used to deliver stimulation pulses as well as to record local field potentials (LFP) from the electrodes. Real-time LFP can be recorded even during periods of electrical stimulation. An external controller Clinician Programmer is used to provide wireless transmission of stimulation parameters and to receive sensed and recorded electrical brain activity.

■ **FIGURE 9.3** Activa® PC+S neurostimulator. *(Copyright © Medtronic. Reprinted with permission.)*

Figure 9.3 shows the two-channel Activa® PC+S neurostimulator which incorporates an internal battery with an expected battery life of 3−5 years, depending on stimulation demand.

9.4 **ST JUDE MEDICAL BRIO™ DBS**

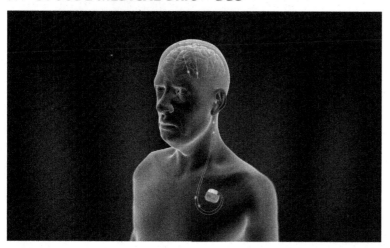

■ **FIGURE 9.4** Brio™ DBS neurostimulator. *(Copyright © St Jude Medical. Reprinted with permission.)*

The Brio™ DBS neurostimulator from St Jude Medical is used for the treatment of Parkinson's disease and primary and secondary dystonia. The Brio™ DBS consists of electrodes implanted in the brain and lead extensions that travel subcutaneously down the neck to the neurostimulator, which is implanted subcutaneously in the upper chest close to the clavicle (Figure 9.4). An external programmer is used to optimize the stimulation parameters and an external patient controller allows the patient to check the battery status and turn the stimulator on and off.

■ **FIGURE 9.5** Brio™ Neurostimulator. *(Copyright © St Jude Medical. Reprinted with permission.)*

Figure 9.5 shows the Brio™ Neurostimulator which incorporates constant current sources for consistent stimulation and bilateral stimulation from one device. The neurostimulator incorporates a high-capacity recharge-able battery which is charged from an external wireless battery charger and can deliver a minimum of 10 years sustained therapy at high parameter settings.

Chapter

Electrical Stimulation Therapy for Epilepsy

10.1 INTRODUCTION

Epilepsy is a disorder of the central nervous system which is related to irregular episodes of electrical activity in the brain, leading to recurrent seizures (fits). These seizures can be classified as a partial seizure, also known as a focal seizure, where a small area of the brain (the focus) is affected, or a generalized seizure, where most of the brain is affected. Partial seizures can be further classified as a simple seizure, where the patient is fully conscious throughout the seizure, or a complex partial seizure, which affects a larger area of the brain and results in confusion and impaired consciousness. Generalized seizures, on the other hand, affect most of the brain, resulting in unconsciousness and uncontrolled muscle convulsions and shaking. A secondarily generalized seizure is where a partial seizure spreads into other areas of the brain and develops into a generalized seizure. Before the onset of a seizure, people may have a sense or aura that a seizure is going to happen. Another form of seizure is status epilepticus which lasts longer than 30 min and can lead to brain damage and can be fatal. When seizures do not respond to drug treatment, the epilepsy disorder is known as refractory or drug-resistant epilepsy and is defined as a failure of adequate trials of two tolerated and appropriately chosen and used antiepileptic drugs (AED) to achieve sustained seizure freedom (Kwan et al., 2010).

The cause of a seizure is not always known and can be affected by physical traumas such as stroke, tumors and head injuries; chemical imbalances brought about by medication, low blood sugar levels, or narcotic drugs; infections such as meningitis; and even flashing lights. Irrespective of the

Implantable Electronic Medical Devices. DOI: http://dx.doi.org/10.1016/B978-0-12-416556-4.00010-3

cause of brain injury, a seizure is related to a disruption or disturbance in the electrical signaling of the neurons, resulting in uncontrollable electrical brain activity, sometimes known as an electric storm. Epilepsy cannot be cured but can be controlled to some extent with medication, a controlled diet, neurosurgery, or electrical stimulation. AED can stop seizures from happening but they do not stop seizures once they have started. A ketogenic diet relies on the body's fat sources to provide energy rather than relying on the breakdown of glucose from carbohydrates. The subsequent production of ketone chemicals resulting from the breakdown of fat in the body has been known to help prevent seizures. A ketogenic diet is essentially a high-fat, low-carbohydrate, controlled-protein diet. Neurosurgery and electrical stimulation can be used to identify areas of the brain responsible for seizures. The seizure focus is known as the epileptogenic locus which can be surgically removed or separated out away from the rest of the brain. Deep brain stimulation may be used to control seizure activity by delivering electrical stimulation via electrodes implanted in a deep brain structure such as the thalamus. It has been shown that electrical stimulation of the anterior nucleus of the thalamus (ANT) can significantly reduce the frequency of seizures (Fisher et al., 2010). Responsive cortical stimulation, where electrical stimulation is delivered to the seizure focus when abnormal electrical brain activity is detected, has also been shown to reduce the frequency of seizures (Morrell et al., 2011).

Vagus nerve stimulation (VNS) can also be used in the treatment of epilepsy in which electrical stimulation pulses are applied to the left vagus nerve in the neck as a means to deliver electrical stimulation into the brain. VNS has been shown to reduce the frequency of seizures but is only used as a supplement alongside prescribed AED.

10.2 SEIZURE-DETECTION METHODS

The effectiveness of reducing seizures is dependent not only on the efficient targeting of the epileptogenic focus or foci but is also dependent on detecting the onset of seizures. Neural brain activity is recorded as electrocorticograms (ECoG) from electrodes placed on or in the brain. When an abnormal ECoG is detected, a stimulus pulse is delivered to disrupt and normalize the abnormal activity. Templates of patient-specific ECoG recordings and waveform detection algorithms can help optimize the analysis and prediction of an epileptic episode before the onset of a seizure, generating the subsequent electrical stimulus response.

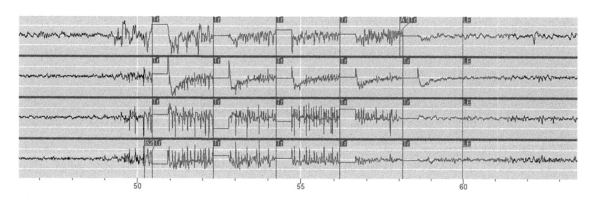

■ **FIGURE 10.1** Abnormal ECoG activity (R2 and A1) normalized after electrical stimulation (Tr). *(Copyright NeuroPace. Reprinted with permission.)*

Electroencephalogram (EEG) recordings from patients suffering from epilepsy show two distinct periods of abnormal brain activity known as the preictal, which represents the onset of a seizure and the ictal region, which is representative of the actual seizure. The preictal period is characterized by distinctive spikes and sharp transient waveforms, whereas the ictal period shows abnormally fast electrical brain activity. Figure 10.1 shows abnormal ECoG activity detected at B2 and A1 and subsequent electrical stimulation delivered at points Tr with the ECoG returning to baseline post stimulation. During and after stimulation, there is a brief period of the sensing amplifier blanking which is shown as a flat line.

Detecting the preictal spikes may give an early warning of an impending seizure. Subsequently, a number of spike-detection algorithms have been developed to predict the onset of a seizure based upon spike features such as amplitude, duration and slope sharpness. These algorithms mainly consist of sampling a section of the EEG waveform in the time or frequency domain, extracting relevant features and classifying the spike parameters to a priori of known predefined features. Techniques used for automatic seizure detection include: mimetic, template matching, wavelet analysis, and artificial neural networks.

Mimetic techniques such as half-wave methods use peak-detector techniques to break down the EEG waveform into segments between consecutive maxima and minima. From the segments, adjacent and of opposite direction, half-waves are derived from which amplitude and duration features are extracted. A wave consists of two back-to-back half-waves from which the slope feature of the wave can be extracted. By taking the average amplitude of the half-waves, a running average can be determined and used for comparison with spike amplitudes. Other features that can be used for extraction include: area, energy, and power spectral density.

Template matching consists of cross-correlating a section of the EEG with a known preictal EEG template depicting a preictal event and comparing the output with a threshold which if exceeded, indicates a possible seizure detection.

Frequency domain analysis decomposes the EEG waveforms into frequency bands for subsequent analysis and is normally performed using the fast Fourier transform, windowed Fourier transform or discrete wavelet transform.

Artificial neural networks utilize an initial training period, after which the network learns how to detect the parameters and features from an EEG waveform that may indicate the possible onset of a seizure.

10.3 NEUROPACE RNS® STIMULATOR NEUROSTIMULATOR

The NeuroPace RNS® neurostimulator is used for the treatment of partial onset and/or secondarily generalized seizures in adults by detecting specific types of electrical activity in the brain and responding with adaptive levels of electrical stimulation to effectively normalize the electrical activity before the onset of seizure. Prior to implantation, areas of the brain are normally identified using EEG recordings and image-scanning techniques such as MRI to help identify no more than two seizure focus sites in order to determine subsequent electrode sites and stimulation threshold levels, which may induce seizures.

The NeuroPace RNS® stimulator uses three seizure-detection techniques; the half-wave, area, and line-length, all of which are configurable such that the parameters can be adjusted to optimize for performance. The line-length algorithm detects changes in both amplitude and frequency where the line-length is defined as being the average of absolute sample to sample differences within a short-term window which is then compared to a long-term window average. Detection occurs when a positive or negative threshold is exceeded. The area algorithm is similar to identifying changes in energy irrespective of frequency where the area is defined as the average absolute area under the curve for a short-term window, which is then compared to the subsequent average area under the curve for a long-term background window. Detection occurs when a positive or negative threshold is exceeded.

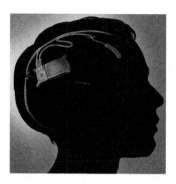

■ **FIGURE 10.2** The NeuroPace RNS® System consists of the RNS Neurostimulator, NeuroPace® Cortical Strip Leads, and NeuroPace® Depth Leads. *(Copyright NeuroPace. Reprinted with permission.)*

Figure 10.2 illustrates the NeuroPace RNS® Neurostimulator which is implanted in the cranium and connected to a NeuroPace® Cortical Strip Lead and a NeuroPace® Depth Lead. The neurostimulator consists of four input channels which can sense continuously and up to 30 min of ECoG activity can be recorded and stored in the neurostimulator at any time. When an abnormal ECoG pattern is detected, the neurostimulator responds by delivering a train of stimulation pulses to the epileptogenic focus to interrupt and normalize the ECoG activity. There are two types of intracranial lead electrodes that can be used (see Figure 10.3), the Cortical Strip Lead, implanted on the surface of the brain and the Depth Lead, implanted within the brain, dependent on the site of the epileptic focus or foci. Each lead consists of four electrodes which are connected to one of the two, four input channels.

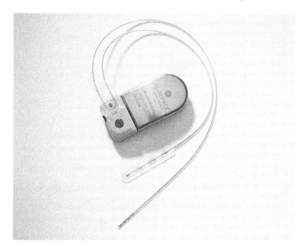

■ **FIGURE 10.3** The NeuroPace RNS® neurostimulator, NeuroPace® Cortical Strip Leads and NeuroPace® Depth Leads. *(Copyright NeuroPace. Reprinted with permission.)*

The neurostimulator shown in Figure 10.3 can be programmed to monitor one or two epileptic foci and can deliver up to five individual sequential stimulation patterns upon detection of a specific ECoG pattern. The stimulation train consists of biphasic current pulses with a programmable pulse width per phase between 40 and 1000 μs, a frequency range of 1–333 Hz and a maximum current output of 12 mA. Each stimulation pattern can consist of two independently configured stimulation bursts lasting from between 10 ms and 5 s. An external programmer wand enables inductive wireless transmission of ECoG data, stimulation parameters, and neurostimulator status. Stimulation can be inhibited by placing an external magnet over the implant and if programmed, a magnet placed over the implant will trigger a recording of ECoG activity. The neurostimulator is battery powered with a 3 V, 705 mAh capacity battery with an expected operating life of around 4 years, dependent on use.

10.4 CYBERONICS INC. VNS

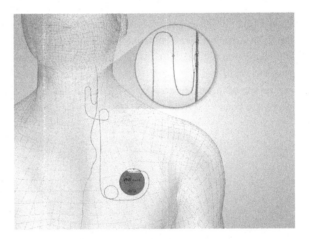

■ **FIGURE 10.4** Cyberonics VNS. *(Copyright © Cyberonics. Reprinted with permission.)*

The Cyberonics VNS consists of an implantable pulse generator, leads, and electrodes, and an external programmer. The stimulator is implanted subcutaneously in the left upper chest region and connects to bipolar electrodes which are attached to the left vagus nerve (Figure 10.4). Stimulation therapy consists of electric current pulses that are sent to the brain via the left vagus nerve at set intervals, night and day. The stimulator incorporates an internal reed switch which is activated by an external magnet to produce on-demand stimulation when patients have a sense or aura to the start of a seizure. The magnet is simply passed over the implant to start the stimulus therapy. Two magnets are supplied; one is a watch-style magnet, attached to a wristband, while the other magnet is a pager-style magnet that attaches to a waistband or belt.

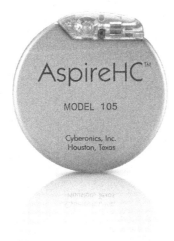

■ **FIGURE 10.5** Cyberonics Therapy Aspire® HC Model 105. *(Copyright © Cyberonics. Reprinted with permission.)*

The pulse generator (Figure 10.5) is implemented using CMOS technology and is powered from an internal lithium carbon monofluoride 3.3 V battery which is stepped up by an internal DC−DC converter to boost the output stage up to a maximum of 12 V. The stimulus pulses consist of charge-balanced biphasic current pulses that can be programmed in 0.25 mA steps up to 3.5 mA with pulse widths ranging from 130 to 1000 μs and a programmable stimulation frequency up to 30 Hz. The battery has a maximum rating of 1.7 Ah such that the stimulator can last for up to 10 years. A programming wand is used to inductively couple data to and from the implanted device.

BIBLIOGRAPHY

Fisher, R., et al., 2010. Electrical stimulation of the anterior nucleus of thalamus for treatment of refractory epilepsy. Epilepsia 51, 899−908.

Kwan, P., et al., 2010. Definition of drug resistant epilepsy: consensus proposal by the ad hoc Task Force of the ILAE commission on therapeutic strategies. Epilepsia 51 (6), 1069–1077.

Morrell, M.J., 2011. RNS system in Epilepsy Study Group. Responsive cortical stimulation for the treatment of medically intractable partial epilepsy. Neurology 77 (13), 1295–1304.

Peripheral Nerve Stimulation

The nervous system of the human body is subdivided into two main areas; the central nervous system (CNS), which is composed of the brain and spinal cord and the peripheral nervous system (PNS) which comprises the nerves that branch from the CNS and innervate areas of the body including the arms, legs and vital organs. The peripheral nerves provide the two-way sensory and motor neural communication between the body and the CNS such that peripheral nerve stimulators can provide some form of functional restoration of muscle movement and sensory pain management therapy.

11.1 DROP FOOT STIMULATORS

11.1.1 Introduction

During a normal gait pattern, the knee extends such that the lower leg swings forward with the foot flexed and raised upward (dorsal flexion) in order to clear the ground. The leg is then extended and the body moves forward, resulting in a heel strike of the foot followed by an extension of the foot (plantar flexion) as the foot is placed down on the floor. This is followed by the "toe off" phase of gait where the heel of the foot starts to raise during the push-off phase of gait (Figure 11.1).

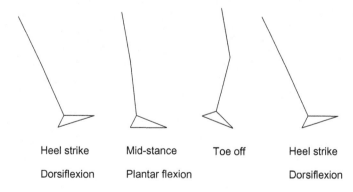

Heel strike   Mid-stance   Toe off   Heel strike

Dorsiflexion   Plantar flexion   Dorsiflexion

■ FIGURE 11.1 Dorsiflexion and plantar flexion phases of gait.

Implantable Electronic Medical Devices. DOI: http://dx.doi.org/10.1016/B978-0-12-416556-4.00011-5

Hemiplegic drop foot is a common cause of gait disruption following a stroke where one side of the body is partially or totally paralyzed. With drop foot, there is no voluntary control of dorsiflexion, resulting in the foot "slapping" on the ground after heel strike and the foot being dragged along during the swing phase of gait. Dorsiflexion can be achieved by electrically stimulating the nerve innervating the tibialis muscle which is the major muscle responsible for dorsal flexion or by stimulating the common peroneal nerve, which evokes the dorsal flexion response and results in knee and foot flexion, raising the foot off the ground. Electrical nerve stimulation can also result in a rotation (eversion) of the ankle when the foot undergoes dorsiflexion. Subsequently, some stimulators provide an extra channel to stimulate the superficial branch of the peroneal nerve to provide for a correction rotation of the ankle. Table 11.1 presents the nerve innervations and muscle groups used for correction of drop foot.

The timing of the electrical stimulation for dorsal flexion which occurs soon after heel strike can be achieved by incorporating a foot switch in the sole of the shoe to detect when heel strike occurs, deactivating the electrical stimulation for dorsal flexion and resulting in the foot to drop naturally. On toe-off, pressure on the heel switch is released, activating the stimulator and evoking dorsiflexion of the foot during the swing

Table 11.1 Nerve Innervations and Muscle Groups Used for Correction of Drop Foot

Nerve	Muscle	Action
Deep peroneal nerve	Tibialis anterior	Dorsiflexes foot
		Inversion of foot
	Extensor hallucis longus	Extends big toe
		Dorsiflexes foot
		Inversion of foot
	Extensor digitorum longus	Dorsiflexes foot
		Extends toes
	Peroneus tertius	Dorsiflexes foot
		Eversion of foot
	Extensor digitorum brevis	Extends toes
	Extensor hallucis brevis	Extends big toe
Superficial peroneal nerve	Peroneus longus	Plantar flexes foot
		Eversion of foot
	Peroneus brevis	Plantar flexes foot
		Eversion of foot

phase of gait. Alternative methods to detect heel strike include acceler-ometers to detect the swing phase of gait and implanted sensors on the sural nerve to detect sensory nerve activity relating to heel strike.

11.1.2 **STIMuSTEP® Finetech Medical Ltd.**

The STIMuSTEP® is used to help correct for drop foot by applying elec-trical stimulation pulses to the nerves that innervate the paralyzed mus-cles required for ankle dorsiflexion and eversion.

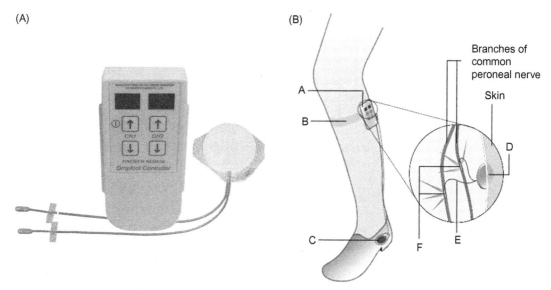

■ **FIGURE 11.2** STIMuSTEP® drop foot stimulator: (A) hardware system and (B) application. *(Copyright © Finetech Medical Ltd. Reprinted with permission.)*

The STIMuSTEP® in Figure 11.2 consists of an implanted receiver (D), a cable made from platinum-iridium wire (E) and two pairs of bipolar epi-neural wire electrodes each inserted into epineurium of the deep branch (F) and the superficial branch of the common peroneal nerve, respectively. A foot switch (C) placed in the shoe under the heel is connected to an external controller (A) which is strapped to the lower leg (B). Power and control signals are inductively transmitted from the external controller to the receiver via the receiver-transmitter coils.

At the start of the swing phase of gait, the foot switch detects heel lift and signals the external controller to transmit power and control signals to the receiver, which subsequently delivers a pattern of electrical stimulation pulses to the two bipolar nerve electrodes to evoke dorsal flexion and rotation of the foot. The stimulation consists of monophasic pulses with a stimulation rate of 30 Hz.

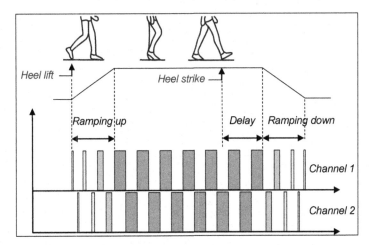

■ **FIGURE 11.3** Stimulation pattern for drop foot. *(Copyright © Finetech Medical Ltd. Reprinted with permission.)*

Figure 11.3 shows the stimulation pattern for drop foot which consists of increasing stimulus pulse widths, ramping up after heel lift and then ramping down after a programmed delay period after heel strike. The stimulation pulses for the two channels are interleaved and complementary in that one bipolar electrode is always active at any time during the swing phase of gait. Table 11.2 provides the adjustable programmable stimulation parameters for the implant.

11.1.3 **ActiGait®, Ottobock**

The ActiGait® is used for the correction of drop foot or for weak dorsiflexion that can arise from a stroke, leading to a partial or complete paralysis of the muscles involved in dorsiflexion.

The ActiGait® in Figure 11.4 consists of a four-channel implantable stimulator which is implanted subcutaneously in the thigh and connects to a cuff electrode which is implanted around the common peroneal nerve sited above the popliteal fossa of the knee. The cuff electrode consists of 12 platinum-iridium nerve contact disks arranged as four sets of three contacts which are connected to the four individual channels of the stimulator

Table 11.2 Adjustable Stimulation Parameters

Parameter	Display	Range	Units	Description	Default
Output power level	Po	1–2	arbitrary	Selects supply voltage to controller	1
Heel-strike time-out (T_{hsto})	Ht	0.1–5.0	0.1 s	Time before ramp down in the event of no heel-strike	3
Ramp up time 1 (T_{ru1})	u1	0.0–5.0	0.1 s	Time for stimulation to ramp up to preset level on channel 1	0.5
Ramp up time 2 (T_{ru2})	u2	0.0–5.0	0.1 s	Time for stimulation to ramp up to preset level on channel 2	0.5
Ramp down time 1 (T_{rd1})	d1	0.0–5.0	0.1 s	Time for stimulation to ramp down to zero on channel 1	0.2
Ramp down time 2 (T_{rd2})	d2	0.0–5.0	0.1 s	Time for stimulation to ramp down to zero on channel 2	0.2
Extension time 1 (T_{ext1})	E1	0.0–5.0	0.1 s	Time delay between heel-strike and start of ramp down for channel 1	0.3
Extension time 2 (T_{ext2})	E2	0.0–5.0	0.1 s	Time delay between heel-strike and start of ramp down for channel 2	0.3
Idle time-out (normal mode)	tO	01–60	minutes	Auto power-off period for Normal Mode	60
Idle time-out (setup modes)	tS	01–10	minutes	Auto power-off period for level mode and parameter mode	10

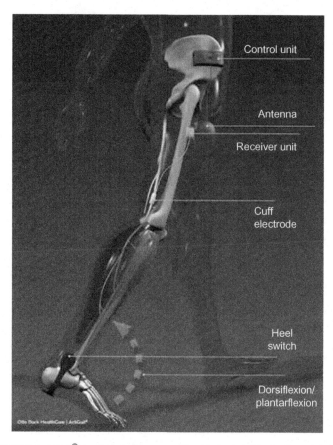

■ FIGURE 11.4 ActiGait® drop foot stimulator system. *(Copyright Ottobock. Printed with permission.)*

(Figure 11.5). The electrode arrangement makes it possible to stimulate different nerve populations within the common peroneal nerve to provide selective activation of the tibialis and peroneus muscle groups required for dorsiflexion and balanced eversion/inversion of the ankle joint.

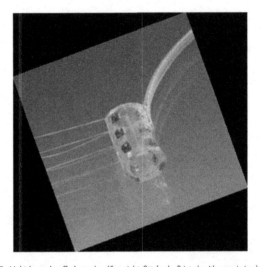

■ **FIGURE 11.5** Multichannel cuff electrode. *(Copyright Ottobock. Printed with permission.)*

An antenna is attached to the skin above the implantable stimulator and connects via an external lead to the controller which can clip onto a belt worn by the user. The heel switch is fastened into a proprietary sock and placed under the foot inside the shoe and wirelessly transmits a signal to the controller during the swing phase of gait to initiate stimulation of the peroneal nerve, thus affecting dorsiflexion of the foot. The heel switch is powered by its own battery with an expected battery life of 1 year whereas the controller is fully rechargeable.

11.2 HANDGRIP STIMULATORS

Stroke and spinal-cord-injured patients, especially regarding an upper motor region, can lose hand function, resulting in loss of handgrip. This is attributed to the partial or total loss of neural activation of the extensor muscles in the forearm essential for wrist and finger extensions. In grip, the fingers are released by an extension of the fingers and wrist.

There are three main nerves innervating the hand; the radial, ulnar and median nerves. One branch of the radial nerve, the posterior interosseous nerve, innervates the extensor muscles in the forearm, providing extensions of the wrist, fingers, and thumb. Table 11.3 presents the main motor nerve innervations for hand function.

Table 11.3 Motor Nerve Innervations and Muscle Groups for Hand Function

Nerve	Muscle	Action
Posterior interosseous nerve	Extensor carpi ulnaris Extensor digiti minimi Extensor digitorum Extensor indicis Extensor carpi radialis brevis Extensor carpi radialis longus	Extends wrists, fingers, and thumb
Ulnar	Interosseous muscles	Spreads fingers apart Bends fingers at knuckles
	Adductor pollicis muscle	Moves thumb back toward hand
Median	Thenar	Opposition of thumb

11.2.1 **STIMuGRIP® Finetech Medical Ltd.**

The STIMuGRIP® uses electrical nerve stimulation to help restore control of wrist extension and subsequent extensions of the fingers to open the hand, restoring function of grip.

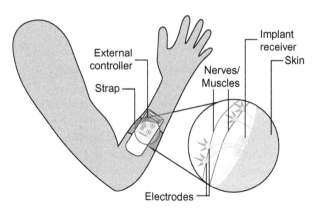

■ **FIGURE 11.6** STIMuGRIP® system. *(Copyright Finetech Medical Ltd. Reprinted with permission.)*

The STIMuGRIP® system consists of a two-channel implanted receiver connected to two pairs of epimysial electrodes, an external controller, and integral three-axis accelerometer attached to the outside of the arm (Figure 11.6). The epimysial electrodes consist of 4-mm-diameter platinum electrodes fixed to an insulating backing to minimize the spread of the electrical stimulation current to other excitable tissue. The electrodes are used in a bipolar configuration with one pair of electrodes placed on the motor point of the extensor carpi radialis brevis muscle to evoke and maintain wrist extension, while the other pair of electrodes are placed on the interosseous nerve to extend the fingers. The controller detects angular displacement of the arm such that different stimulation strategies can be pre-programmed based on the angular movement detected to cater for a wide range of activities. For example, moving the forearm to the horizontal position initiates the stimulation strategy for program 1, which could be wrist extension followed by finger extension and when the arm moves back, the stimulation is turned off and the fingers, due to the natural spasticity of the hand, close around the object of interest. Conversely, moving the arm side to side could initiate program 2 for wrist extension but no subsequent finger extension.

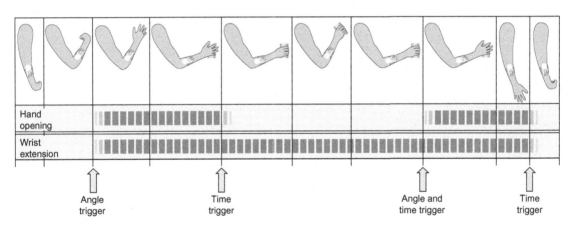

■ **FIGURE 11.7** An example of stimulation cycle to grip and raise a glass beaker. *(Copyright Finetech Medical Ltd. Reprinted with permission.)*

Figure 11.7 illustrates an example of a stimulation cycle in which raising the forearm triggers stimulation for wrist extension and for finger extension to open the hand. After a predefined time interval, the stimulation for finger extension is turned off, causing the hand to close, due to its natural spasticity, around the object which can then be lifted up. When the forearm is lowered, there is a predefined time interval before stimulation is applied to open the hand and release the grip on the object. Lowering the arm further, turns off the stimulator.

Lower Esophagus Stimulator

12.1 **INTRODUCTION**

The stomach is essentially a hollow muscular organ which secretes protein-digesting enzymes and hydrochloric acid to aid in the digestion of food which enters via the esophagus. The thicker layer of muscle tissue surrounding the region where the esophagus meets the stomach forms the lower esophageal sphincter which is usually in a contracted state to block the backflow or reflux, of acid from the stomach into the esophagus. The lower esophageal sphincter normally opens to allow for swallowing, burping and vomiting. Muscular gastric folds inside the stomach allow the stomach to stretch and also aid in digestion by contracting periodically. Partially digested food is then pushed out of the stomach via the pyloric sphincter to the duodenum which forms part of the small intestine.

The collective terms, Gastro-oesphageal Reflux Disease (GORD) also known as Gastro-esophageal Reflux Disease (GERD) identifies the most common gastrointestinal disorders relating to acid reflux. If acid from the stomach leaks into the esophagus, then this acid reflux is normally felt as heartburn, a burning feeling in the chest after eating. Regurgitation is the condition where sour stomach acid is tasted in the mouth. Esophagitis occurs when the lining of the esophagus becomes irritated and inflamed from the stomach acid which can lead to painful sores and ulcers. In severe cases, this can result in Barrett's disease and lead to cancer of the esophagus. In most cases, conditions relating to GERD can be treated with changes in diet, medication such as protein pump inhibitors and in extreme cases, by surgery.

Normally, when food is swallowed, the esophageal sphincter muscle relaxes and then closes by muscle contraction to prevent acid reflux. However, if the esophageal sphincter is not functioning properly, then acid reflux can occur (Figure 12.1).

An alternative treatment for GERD is to electrically stimulate the lower esophageal sphincter muscle to promote sufficient contraction and closure of the sphincter. This effectively blocks acid reflux without interfering

Implantable Electronic Medical Devices. DOI: http://dx.doi.org/10.1016/B978-0-12-416556-4.00012-7

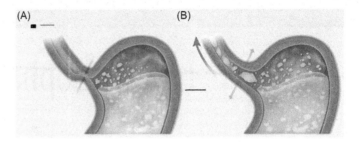

■ **FIGURE 12.1** Esophageal sphincter between esophagus and stomach (A) normally closed and (B) reflux. *(Copyright © EndoStim. Adapted and reprinted with permission.)*

with relaxation of the esophageal sphincter and allows for normal swallowing function.

12.2 ENDOSTIM® LOWER ESOPHAGUS STIMULATOR

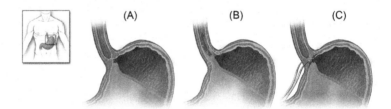

■ **FIGURE 12.2** Lower esophageal sphincter (A) normally closed, (B) reflux, and (C) placement of IM electrodes. *(Copyright © EndoStim. Reprinted with permission.)*

The Lower Esophageal Stimulator (LES) from EndoStim® uses electrical stimulation via implanted intramuscular (IM) electrodes in the lower esophageal sphincter muscle to effectively close the sphincter and prevent reflux but still maintain normal swallowing function. Figure 12.2 shows the sphincter normally closed in (A) and stomach reflux in (B), whereas (C) shows the placement of the IM electrodes in the lower esophageal sphincter muscle to effect a contraction and hence closure of the sphincter.

The LES system consists of an Implantable Pulse Generator (IPG) and, a bipolar electrode lead with a pair of platinum-iridium bipolar stitch electrodes at the end. The first-generation LES IPG has an estimated battery life of greater than 10 years operating with nominal parameters, while the smaller second-generation IPG, has an estimated battery life of greater than 7 years operating at nominal parameters (Figure 12.3).

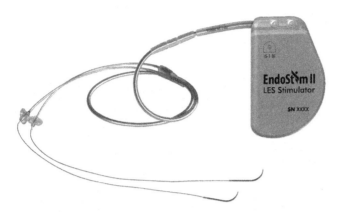

■ **FIGURE 12.3** Lower Esophageal Stimulator. *(Copyright © EndoStim. Reprinted with permission.)*

The electrodes are surgically implanted by conventional laparoscopy and endoscopic visualization of the gastroesophageal junction. The electrodes are implanted in the muscular layer in the right anterior quadrant of the esophagus (Figure 12.4) and secured in place with silicone butterfly tabs. The electrode leads are delivered through the abdominal wall and connected to the IPG which is implanted in a subcutaneous pocket in the left upper quadrant of the abdomen. Following implantation, the IPG is interrogated and programmed by a wireless programmer wand placed over the IPG site.

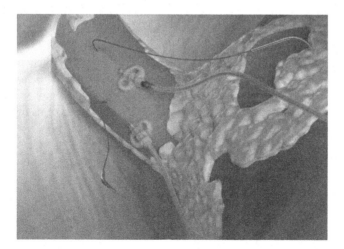

■ **FIGURE 12.4** Bipolar stitch electrodes inserted into the musculature of the lower esophageal sphincter. *(Copyright © EndoStim. Reprinted with permission.)*

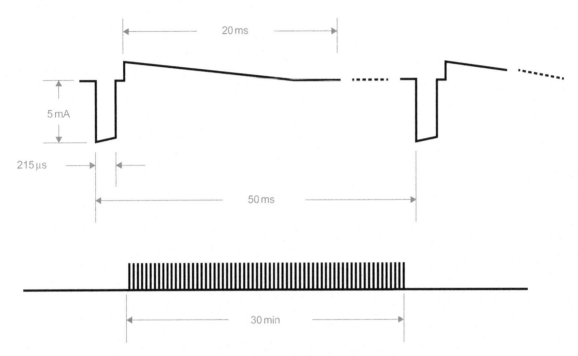

■ **FIGURE 12.5** Applied stimulus pulse pattern. *(Copyright © EndoStim. Reprinted with permission.)*

The applied stimulus pulse consists of a 215-µs wide negative phase monophasic pulse with a nominal amplitude of 5 mA (range 3−8 mA) followed by a 20 ms charge balancing phase with a maximum amplitude of 1 mA (Figure 12.5). The stimulus is applied at a rate of 20 Hz over a 30 min period for each treatment session and is repeated up to 12 times a day and delivered premeal and pre-flux, based on patient's symptoms and baseline 24 h pH recordings. The stimulation can be optimized to meet the individual patient's needs by adjusting the number and timing of sessions and stimulation parameters to address residual symptoms, suboptimal symptoms or residual acid events on pH testing. The IPG also incorporates an accelerometer to detect upright and supine states such that the stimulation algorithm can be customized, based on a patient's position to address supine and nocturnal reflux.

Chapter

Vagal Blocking Therapy

13.1 INTRODUCTION

The vagus nerve, also known as the tenth cranial nerve, originates in the medulla region of the brain stem. It consists of a pair of nerves which contains both sensory and motor nerve fibers with attachments to the internal organs relating to a multitude of biological functions such as breathing, digestion, metabolism and heart rate. In the case of digestion, it has been shown that the vagus nerve plays an important part in gastrointestinal activity such as the secretion of gastric acid and digestive enzymes. The vagus nerve also transmits neural information relating to the feeling of hunger (satiety) and fullness (satiation), as well as controlling energy metabolism essential for digestion. Subsequently, if the sensory mechanism related to the feeling of hunger can be suppressed, then this can theoretically reduce the desire to consume food and may help in the treatment of obesity. Consequently, blocking the neural transmission of the sensory afferent neural signals from the digestive system using electrical signals, it may be possible to reduce the feeling of hunger and help in the treatment of obesity.

13.2 ENTEROMEDICS® VBLOC VAGAL BLOCKING THERAPY

Vagal nerve activity is blocked throughout the waking hours of the patient, on average 12 h a day, using proprietary VBLOC® Therapy delivered using the Maestro® System. VBLOC® Therapy uses intermittent high-frequency, low-energy, electrical pulses to block the transmission of naturally occurring sensory signals that travel on the two trunks of the vagus nerve from the stomach to the brain and vice versa to suppress the sensation of hunger and promote the feeling of fullness. An electrical block of the vagus nerve can also affect the release of digestive enzymes and can regulate the muscular contractions of the stomach to promote slow emptying of the stomach contents, all of which may reduce the desire to consume food, which may ultimately lead to weight loss.

Implantable Electronic Medical Devices. DOI: http://dx.doi.org/10.1016/B978-0-12-416556-4.00013-9

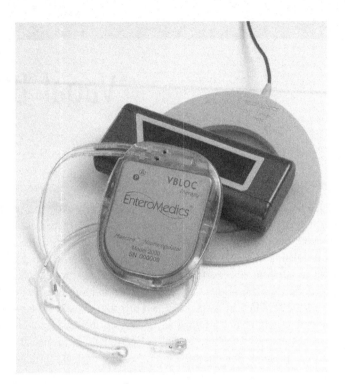

■ **FIGURE 13.1** EnteroMedics Maestro® System vagal blocking device. *(Copyright EnteroMedics. Reprinted with permission.)*

Figure 13.1 shows the Maestro® System which consists of an implantable device known as the Neuroregulator, proprietary leads and electrodes, a mobile charger, and a transmit coil. The Neuroregulator is surgically implanted subcutaneously just below the ribs on one side of the body. The two electrodes are laparoscopically placed on the two vagus nerve trunks just above the junction between the esophagus and the stomach. The mobile charger and transmit coil are worn externally for a short time daily by the patient to charge the battery in the implanted Neuroregulator. An external Clinician Programmer is used to set up the required therapy parameters as well as to store data regarding system performance.

Implantable Drug Delivery Systems

14.1 INTRODUCTION

Implantable drug delivery systems (DDS) are designed to store and deliver small, precise doses of therapeutic drugs or medicines into the blood stream or to specific tissue sites, subsequently replacing the daily injection of drugs required for pain relief and for the treatment of many conditions and diseases such as osteoporosis, heart disease, cystic fibrosis, glaucoma, age-related macular degeneration (AMD), diabetes, refractory epilepsy, and cancer. With the advent of microtechnology and nanotechnology the components of DDS can be fabricated on a small scale, allowing for miniature systems to be implanted locally to the drug delivery site, for example, in the eye or in the intrathecal space surrounding the spinal cord.

Implantable DDS essentially consist of a micropump that contains a reservoir in which the pharmaceutical drug in gaseous or liquid form is stored, an actuator release or pump mechanism, inlet and outlet valves, and in some cases a cannula or catheter to direct delivery of the drug to a target site. The drug is released using either a passive or an active (responsive) micropump in the DDS. A typical passive actuator relies on a drug infusion process in which the drug is slowly released through a porous membrane or a biodegradable membrane that degrades over time. Other passive methods include an array of small sealed drug wells that make up the reservoir such that dissolving the individual seals by applying an electrical stimulus releases a precise drug dose. In either case, the released drug diffuses into the target delivery site.

Because of the diverse range of active micropumps, a number of classifications have grouped micropumps into mechanical (reciprocating) and nonmechanical (no moving parts) pumps. However, micropumps can be grouped into displacement and dynamic pumps (Krutzch and Cooper, 2001). Displacement pumps exert pressure forces on the working fluid through one or more moving boundaries and dynamic pumps continuously add energy to the working fluid in a manner that increases either its momentum or its pressure directly (Laser and Santiago, 2004).

Implantable Electronic Medical Devices. DOI: http://dx.doi.org/10.1016/B978-0-12-416556-4.00020-6

Essentially, active micropumps respond to an electrical stimulus to cause an electrochemical or electromechanical response to affect a controlled drug release that can be continued or stopped at any time. Many forms of micropumps have been designed and implemented using actuators based on processes such as electrolysis, osmosis, hydrodynamics, electrophoresis, piezoelectrics, magnetics, pneumatics, hydrolysis, and material deformations.

14.2 **ELECTROMAGNETIC MICROPUMPS**

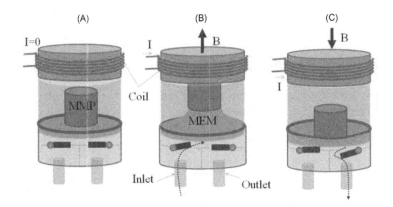

■ **Figure 14.1** Pictorial representation of an electromagnetic micropump: (A) close mode, (B) suction mode, and (C) pumping mode (MEM, membrane, MMP, magnetic moving part, I, electric current, B, magnetic field (Hamie et al., 2013)). *(Reprinted with permission.)*

Electromagnetic micropumps consist of a fluid chamber (reservoir) that has a coil of wire at one end surrounding a magnet fixed to a flexible membrane such as polydimethylsiloxane (PDMS) and inlet and outlet valves placed at the other end. Applying a current to the electromagnetic coil, and depending on the direction of current flow, causes the magnet to either push down or pull up on the membrane, causing a change in chamber pressure such that fluid is either drawn into or pumped out of the reservoir and is directed to the drug release orifice outlet, usually through one-way valves (Figure 14.1). The required drug dose and delivery rate are controlled by the magnitude and duration of the applied electric current. A similar approach is to embed ferromagnetic pieces into the membrane such that the application of a magnetic field will cause displacement of the membrane (Khoo and Liu, 2000).

14.3 **OSMOTIC MICROPUMPS**

Two solutions of different solute concentrations separated by a semiper-meable membrane will establish osmotic pressure across the membrane such that there will be a net flow of water (the solvent) through the membrane from the low-concentration solution to the high-concentration solution. An equilibrium state will be reached when the difference in pressure between the two solutions is equal to the osmotic pressure.

The main components of an osmotic micropump include the pharmaceutical drug (solute), an osmotic agent, a semipermeable membrane, a solvent (usually water), and a delivery orifice. Osmotic agents also known as osmogens are used to increase the osmotic pressure, which is directly proportional to the drug delivery rate.

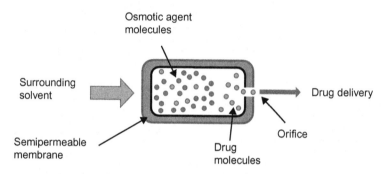

■ **Figure 14.2** One-compartment osmotic micropump.

Osmotic micropumps can be classified into three types based on the com-partmental separation of the pharmaceutical drug, the osmotic agent, and the solvent, which is normally water derived from surrounding aqueous body tissue fluids. In its simplest form, an osmotic pump consists of a single compartment in which the drug is separated from the solvent that flows through a semipermeable membrane to dissolve the drug (Figure 14.2). The drug acts as the osmotic agent and its solubility deter-mines the drug release rate, which is kept relatively constant as long as the drug solution remains in saturation.

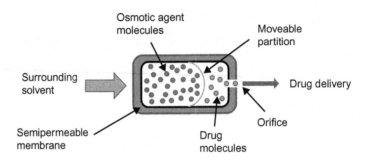

■ Figure 14.3 Two-compartment osmotic micropump.

In a two-compartment osmotic micropump, an osmotic agent is dissolved by water derived from surrounding aqueous body tissue fluids to create the osmotic pressure. The micropump consists of two compartments or reservoirs, a drug reservoir and an osmotic agent reservoir both separated by a moveable barrier or partition. When the osmotic agent dissolves, the resultant pressure of the saturated osmotic agent pushes against the barrier, displacing the drug solution toward the orifice of the micropump. An alternative implementation is to have a flexible membrane wall for the osmotic agent reservoir that can expand or swell under pressure, thus exerting pressure on the drug solution to flow through the orifice of the micropump (Figure 14.3).

Micropumps can also be implemented having three or more compartments. Three-compartment micropumps are an extension of two compartments in which the extra compartment holds the solvent rather than relying on aqueous body tissue fluids to act as the solvent.

The fundamental equation for all osmotically driven pumps (Simon et al., 2012) is given by:

$$\frac{dm}{dt} = KA\pi C \tag{14.1}$$

where:

$\dfrac{dm}{dt}$—drug release rate of the mass of drug molecules released over time through the outlet orifice of the osmotic pump

K—permeability of the semipermeable membrane

A—surface area of the semipermeable membrane

C—drug concentration of the dispensed solution

14.4 **ELECTRO-OSMOTIC MICROPUMPS**

When a liquid comes into contact with the surface of a solid material, the ions in the liquid interact with the surface of the solid, forming a single charged layer. Ions of opposite charge in the liquid are then attracted to the first layer to neutralize the resultant charge and subsequently form a second or electric double layer (EDL) of charges. Because the ions in the second layer are loosely bound to the first layer, the application of an electric field can cause migration of these loose ions to effectively drag the bulk of the liquid along by interaction of viscous forces.

Using a direct current (DC) to generate the electric field can result in a linear flow response whereby the resultant velocity of the liquid is proportional to the charge density in the EDL and the strength of the applied electric field. However, because the field electrodes are in contact with the liquid, there is the risk of electrolysis at the electrodes, especially with water, which will produce bubbles of oxygen and hydrogen gases that can impede the flow of the liquid.

Using an alternating current (AC) to generate the electric field can minimize the risk of electrolysis but the net liquid flow will be zero. Applying the same field current to pairs of asymmetrical electrodes placed along the solid walls of the micropump results in a net flow of fluid from the smaller to the larger electrodes. The flow of liquid depends on many factors such as the applied electric field, the electrode geometry, and the number of electrodes. Because the fluid flow can be controlled, electro-osmotic pumps do not incorporate valves.

14.5 **ELECTROLYSIS MICROPUMPS**

Electrolysis micropumps involve the electrolysis of an electrolyte into gases that ultimately provide the pneumatic pressure of the micropump to pump the pharmaceutical drug out of the drug reservoir into a cannula for delivery to the drug site. The electrolyte normally used is water which decomposes into oxygen and hydrogen in a chamber separate from the drug reservoir. However, a pharmaceutical drug in aqueous form can be used as the electrolyte for the decomposition of water.

Electrolysis pumps use noble metal electrodes such as platinum placed in an aqueous solution such that passing an electric current through water, results in the decomposition of water into oxygen and hydrogen gases. At the cathode:

$$2H_2O \leftrightarrow 4H^+ + 4e^- + O_2 \qquad (14.2)$$

At the anode:

$$4H^+ + 4e^- \leftrightarrow 2H_2 \qquad (14.3)$$

resulting in:

$$2H_2O \leftrightarrow 2H_2 + O_2 \qquad (14.4)$$

which is a reversible reaction such that removing the applied current, results in a gradual recombination of the oxygen and hydrogen gases to water.

The rate of gas produced is given by:

$$\frac{\Delta V}{\Delta t} = \frac{3}{4} \frac{i}{F} Vm \qquad (14.5)$$

where:

ΔV—total amount of gas produced (m^3)
Δt—time duration of the applied current (s)
i—applied current (A)
F—Faraday's constant (96.49×10^{-3}°C/mol)
Vm—molar gas constant at 25°C and atmospheric pressure of
    24.7×10^{-3} m^3/mol

Subsequently, the required pharmaceutical drug dose and delivery rate can be controlled by the magnitude and duration of the applied electric current. Using an interdigitated (finger) electrode array helps minimize currents paths through the solution that would otherwise cause heating effects in the solution and improves the efficiency of the micropump.

14.6 WIRELESS MICROCHIP DRUG DELIVERY SYSTEM BY MicroCHIPS INC.

MicroCHIPS has a number of DDS in development for the treatment of osteoporosis, diabetes, and multiple sclerosis, and a remote control contraception implant (Farra et al., 2012). Contraceptive implants are normally passive in that they provide a slow release of a progestogen hormone such as etonogestrel and levonorgestrel that stops the release of eggs from the ovaries. The MicroCHIPS DDS contraceptive implant is an active system in that it contains microreservoirs of 30 µg of the hormone levonorgestrel released daily over a 16 year period and the process can be deactivated and reactivated at any time using a short-range wireless remote control unit to minimize accidental interference from external sources.

The MicroCHIPS implant is also used for the treatment of osteoporosis, which is a progressive disease that affects the density of bone causing bones to become weak and fragile. If left untreated, osteoporosis ultimately results in fracture. One treatment for osteoporosis in which the bone density can be increased includes the administration of the human parathyroid hormone (PTH) teriparatide, PTH(1-34), which is injected into the body tissue on a daily basis. The hormone is used to stimulate the manufacture of new bone osteoblasts in the body to increase bone mass. Subsequently, the MicroCHIPS DDS for the treatment of osteoporosis includes reservoirs of PTH(1-34) that are released on a daily basis.

(A)   (B)

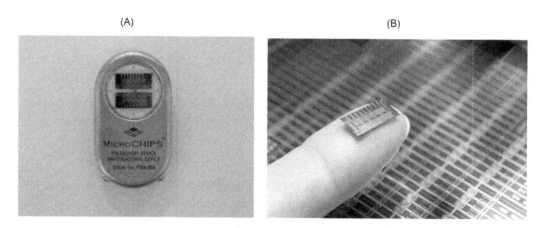

■ **Figure 14.4** (A) MicroCHIPS implantable drug delivery system. (B) MicroCHIPS array of reservoirs. *(Copyright © MicroCHIPS, Inc. Reprinted with permission.)*

The MicroCHIPS DDS shown in Figure 14.4A is based on the microfabrication of multiple wells or reservoirs (Figure 14.4B) in a microchip that forms part of an implantable device that allows for multiple drugs to reside within the same chip and for drug doses to be released in any order. The DDS also incorporates a real-time clock and wireless telemetry circuits for active control of drug dose scheduling from an external remote control unit placed in contact with the skin over the site of implantation. The bidirectional communications link also provides information regarding dose delivery confirmation and implant status such as battery voltage.

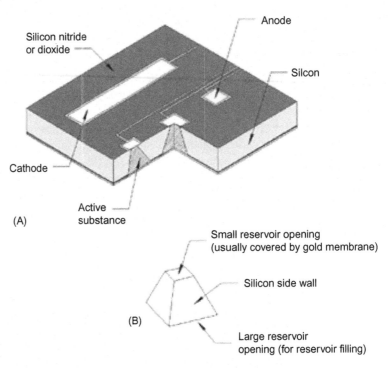

■ Figure 14.5 MicroCHIPS structure showing (A) the pyramidal wells and (B) electrode contacts. *(Copyright © MicroCHIPS (Santini Jr, et al., 1999). Reprinted with permission.)*

The MicroCHIPS structure shown in Figure 14.5 consists of a silicon substrate into which are etched pyramidal microreservoirs that are filled with the prescribed drug dose and then hermetically sealed with an ultra thin gold membrane covering that also serves as an anode electrode. Applying an anodal DC voltage of 1.04 V with reference to the cathode for a period of several seconds, results in the gold membrane to effectively dissolve (Figure 14.6) such that the drug in the reservoir can diffuse into the surrounding tissue and subsequently enter the blood stream. An alternative implementation is to use thin metal films that effectively act as electrical fuses when heated to the point of failure (blown) when a predetermined current is passed through the film.

The MicroCHIPS measures 20 mm × 20 mm × 7 mm and is designed to be implanted subcutaneously in the buttocks, upper arm, or abdomen under local anesthetic.

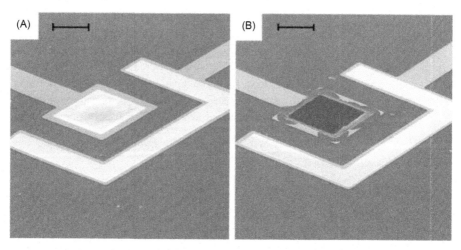

■ **Figure 14.6** Scanning electron micrographs of anodal gold membrane covering before (A) and after (B) the applied anodal voltage. *(Copyright © MicroCHIPS (Santini Jr, et al., 1999).)*

14.7 CODMAN® 3000 CONSTANT FLOW INFUSION SYSTEM IMPLANTABLE PUMP BY CODMAN & SHURTLEFF, INC.

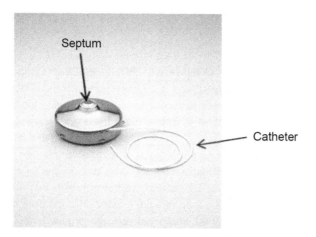

■ **Figure 14.7** CODMAN® 3000 Constant Flow Infusion System Implantable Pump. *(Copyright © Codman & Shurtleff, Inc. Reprinted with permission.)*

The CODMAN® 3000 Implantable infusion drug pump shown in Figure 14.7 is an example of a passive DDS used to deliver medications at a constant rate for the treatment of chronic pain, severe spasticity and cancers of the liver. The CODMAN® 3000 does not require batteries and is available in three sizes (16, 30, and 50 mL) to increase the delivery time between refills. The FLEXTIP Plus SureStream intraspinal catheter is used when medication is delivered to the intraspinal space as required in the treatment for chronic pain and severe spasticity. The drug reservoir is refilled by inserting a needle through the skin into the central septum through a self-sealing silicone membrane.

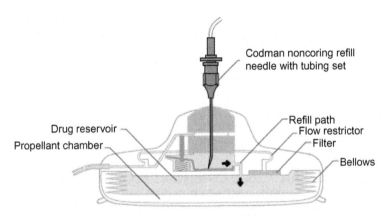

Codman noncoring refill needle with tubing set

Refill path
Flow restrictor
Filter
Bellows

Drug reservoir
Propellant chamber

■ **Figure 14.8** Cross-sectional view of the CODMAN® 3000 Pump. *(Copyright © Codman & Shurtleff, Inc. Reprinted with permission.)*

Figure 14.8 shows the CODMAN® 3000 pump that contains two chambers, the inner drug reservoir and the permanently sealed outer propellant bellows chamber that contains a compressible fluorocarbon gas. When the drug reservoir is empty, the bellows contract under pressure from the gas in the propellant chamber which is dependent on the body temperature. Filling or refilling the drug reservoir chamber causes the bellows to expand, which forces the drug to flow out of the drug reservoir through a filter and a capillary restrictor tube which maintains an optimal flow rate of the drug through a catheter to the drug delivery site. The CODMAN® 3000 Pump delivers the required medication, with typical flow rates between 0.5 and 2 mL/day. In cases in which a direct injection into the body is required or when the catheter needs flushing with saline, the CODMAN® 3000 Bolus Needle is used which makes a direct connection to the tip of the catheter. The CODMAN® Pump 3000 is housed in a titanium casing and is implanted subcutaneously in the lower abdomen.

14.8 **SYNCHROMED® II INFUSION SYSTEM BY MEDTRONIC**

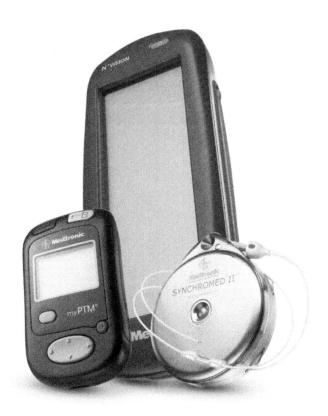

■ **Figure 14.9** SynchroMed® II Infusion System consisting of the myPTM, N'Vision Clinician Programmer, and the SynchroMed® II Programmable Infusion Pump. *(Copyright © Medtronic. Reprinted with permission.)*

The SynchroMed® II Infusion System delivers pain medication directly to the intrathecal space surrounding the spinal cord. The treatments include pain management for severe chronic nonmalignant pain (CNMP), severe spasticity of spinal or cerebral origin, and treatment of primary or metastatic cancer. Figure 14.9 shows the SynchroMed® II Infusion System, which consists of the myPTM (Personal Therapy Manager), the N'Vision Clinician Programmer and the SynchroMed® II Programmable Infusion Pump.

The myPTM is a handheld personal therapy manager that gives patients the control to receive extra prescribed bolus doses of medication when the need arises. The N'Vision Clinician Programmer is used to create and store up to 50 clinician-created programs for the management of patient treatment, as well as to store all critical therapy and patient information.

The SynchroMed® II Programmable Infusion Pump is implanted in the abdominal area and is connected to a small thin catheter that is implanted in the intrathecal space. Two sizes of drug reservoirs are available, 18 and 20 mL. An alarm is activated if the drug level decreases to less than 2 mL. The pump can be programmed to deliver medication at therapeutic flow rates from 0.002 to 1.0 mL/h (0.048–24 mL/day). The battery life, depending on usage, is between 4 and 7 years. After 7 years, the pump automatically shuts off.

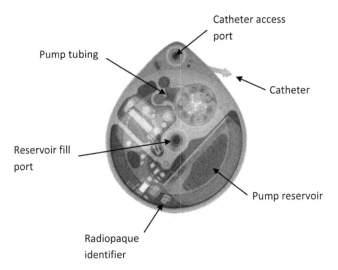

■ **Figure 14.10** SynchroMed® II Programmable Pump. *(Copyright © Medtronic. Reprinted with permission.)*

Figure 14.10 shows the internal components of the SynchroMed® II Programmable Pump. The drug is delivered using a 22-gauge corning needle into the reservoir fill port, where it passes through the reservoir valve and into the pump reservoir. The micropump contains a sealant gas that begins to heat up from the body's temperature, exerting pressure on the pump reservoir forcing the drug into the pump tubing, where an electronically controlled motor pushes the required dose out through the catheter port and into the catheter to the infusion site. The catheter access port allows for direct access to the catheter tip using a 24-gauge noncoring needle. This allows for direct access to the infusion site and also for flushing of the catheter.

The radio-opaque identifier is used to record the manufacturer and model code, which are visible using standard X-ray procedures.

14.9 **MIP IMPLANTABLE FROM DEBIOTECH**™

■ **Figure 14.11** Two faces of the MIP Implantable micropump. *(Copyright © DebioStar™. Reprinted with permission.)*

The MIP Implantable pump from Debiotech is a piezo-actuated silicon micropump fabricated using MEMS technology working as a volumetric pump by the reciprocating action of a silicon micromachined membrane to periodically compress the fluidic chamber to pump the pharmaceutical drug out through one-way directional fabricated valves. Figure 14.11 shows two faces of the MIP micropump.

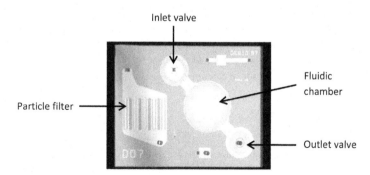

■ **Figure 14.12** Intermediate silicon layer showing the micromachined circular fluidic chamber and inlet and outlet valves. *(Copyright © Debiotech™. Reprinted with permission.)*

The silicon layers are micromachined to form the circular fluidic chamber and valve structures (Figure 14.12), whereas the glass layers contain through holes for the fluid. On top of the stack is a piezoelectric ceramic disc that is bonded to the top silicon layer, which collectively acts as the micropump actuator (Figure 14.13). At the bottom of the stack are two titanium fluid connectors hermetically joined to the micropump. The filter acts to protect the sensitive elements of the micropump against the intrusion of particles and bacteria.

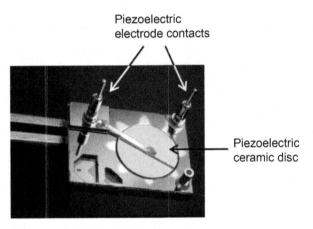

■ **Figure 14.13** MIP micropump top layer showing the piezoelectric ceramic disc.
(Copyright © Debiotech™. Reprinted with permission.)

The piezoelectric actuator vibrates when a voltage is applied across the material, initiating a displacement of the silicon membrane, initiating a reciprocating pump action. The structure of the device is such that on the compression stroke, the membrane compresses the fluid chamber and directs the fluid flow through the inlet and outlet valves. The flow rate is linear with the actuation frequency up to 0.2 Hz and achieves a typical flow rate of 0.1 mL/h with a stroke volume of 150 nL. The whole chip measures 16 mm × 12 mm × 1.86 mm.

14.9.1 **DebioStar**™

The DebioStar™ is an implantable DDS that can be used for local and sustained delivery of pharmaceutical drugs subcutaneously, intraperitoneally or intramuscularly for single or refillable use. Figure 14.14 shows the DebioStar™ that contains a refillable drug reservoir and a controlled silicon nanoporous membrane, whereby the pore diameter and membrane thickness

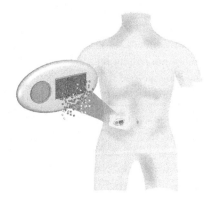

■ **Figure 14.14** DebioStar™ system consists of a refillable reservoir and a controlled nanoporous membrane. *(Copyright © DebioStar™. Reprinted with permission.)*

can be controlled to alter the drug delivery rate such that the drug can be delivered over several weeks or several months. The pore diameters can be altered with a range between 1 and 250 nm and the total membrane range thickness is from 50 nm to several hundred micrometres. A pore density of 1 billion pores/cm^2 is possible. Also, by chemically altering the membrane surface properties, the drug delivery can be delayed over longer periods.

14.10 OPHTHALMIC MICROPUMP™ REPLENISH, INC.

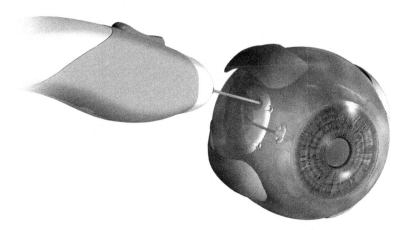

■ **Figure 14.15** Ophthalmic MicroPump System™. *(Copyright © Replenish, Inc. Reprinted with permission.)*

The Ophthalmic MicroPump System™ has been developed to provide drug delivery therapy for patients diagnosed with glaucoma or retinal pathologies who would normally rely on frequently administered drug therapy such as monthly intravitreal injections. The MicroPump™ system (Figure 14.15) can be programmed to deliver a personalized therapy consisting of microliter-sized doses every hour, day, or month according to patient's need.

The Anterior MicroPump™ has been designed for glaucoma patients and incorporates a cannula system that is inserted into the anterior segment of the eye. The Posterior MicroPump™ has a Pars Plana Clip™ that allows drug to be delivered into the posterior segment of the eye in patients with retinal disease. The MicroPumps have a typical refill time interval of several months and are refillable using a 31-gauge needle connected to the Drug Refill System™. In addition, a wireless telemetry based system is used to transmit data from the device to a separate EyeLink™ console that communicates with and recharges the MicroPump.

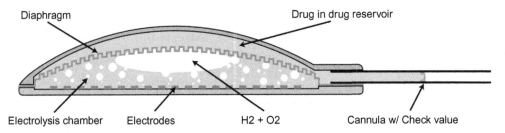

Diaphragm   Drug in drug reservoir

Electrolysis chamber   Electrodes   H2 + O2   Cannula w/ Check value

■ **Figure 14.16** Diagrammatic representation of the pump. Gas bubble evolution resulting from electrolysis in drug reservoir packaged electrically controlled device. *(Copyright © Replenish, Inc. Reprinted with permission.)*

The MicroPumps are fabricated using microelectromechanical systems (MEMS) technology and operate using an electrolysis mechanism powered by either embedded batteries or inductively coupled power coils (Figure 14.16). Water is dissociated into hydrogen and oxygen such that the gas generates sufficient pressure to force the drug through a check valve into the eye. The required drug dose and delivery rate is controlled by the magnitude and duration of the applied electric current.

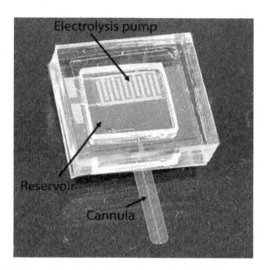

■ **Figure 14.17** Packaged electrically controlled mini drug pump. *(Copyright © Replenish Inc. (Saati, et al., 2009).)*

The MicroPump™ assembly shown in Figure 14.17 consists of a separate drug reservoir integrated on top of the electrolysis pump and incorporates a self-sealing silicone membrane to facilitate the injection of the pharmaceutical drug into the reservoir. The electrolysis pump is fabricated with Parylene using MEMS technology and incorporates platinum interdigitated electrodes and a microchannel parylene cannula. Figure 14.18 shows an artist's rendition of a MicroPump™ currently in development.

■ **Figure 14.18** An artist's rendition of a MicroPump™. currently in development. *(Copyright © Replenish, Inc. Reprinted with permission.)*

14.11 INTELLIDRUG™ SYSTEM FROM INTELLIDRUG

The IntelliDrug™ implant has been designed to fit in the mouth, occupying the space between the molars and the cheeks. From this site, the implant can deliver a controlled release of pharmaceutical drugs directly into the bloodstream via the inside lining of the cheeks (buccal mucosa) which has greater permeability to that of skin. The IntelliDrug™ system shown in Figure 14.17 consists of the micropump implant and a communications network incorporating wireless and RFID technology that allows the required dosage to be set by a remote medical centre. A transdermal link placed between the device placed in the oral cavity and an extra-corporeal communications device is also possible (Figure 14.19).

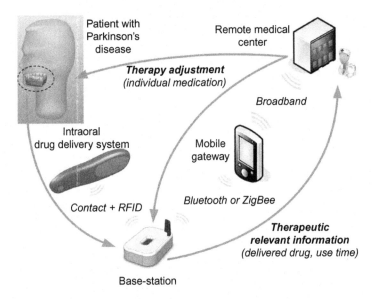

■ **Figure 14.19** IntelliDrug™ system. *(Copyright © IntelliDrug™. Reprinted with permission.)*

The IntelliDrug™ implant is based on an osmotic micropump that consists of a reservoir that holds the drug in a solid pill form, a compressible polymer balloon (fluidic capacity), an electronically controlled microvalve and a flow sensor. Figure 14.20 shows the implant fixed to a mandibular tooth such that water from saliva passes through the semipermeable membrane (saliva inlet) and dissolves the drug, creating a hydrostatic pressure that forces the dissolved drug through the microvalve drug outlet for delivery. The flow sensor uses impedance measurement techniques to sense the drug flow rate and concentration of the dissolved drug solution in order to alert the user when the concentration level drops below a predetermined level.

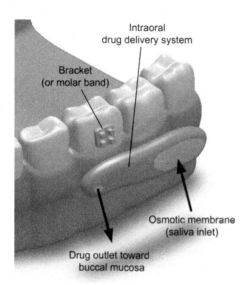

■ Figure 14.20 IntelliDrug micropump. *(Copyright © IntelliDrug. Reprinted with permission.)*

BIBLIOGRAPHY

Farra, R., Sheppard, N.F., McCabe, L., Neer, R.M., Anderson, J.M., Santini, J.T., et al., 2012. First-in-Human testing of a wirelessly controlled drug delivery microchip. Sci. Transl. Med. 4 (122), 122ra21.

Hamie, A.H., Ghafar-Zhadeh, E., Sawan, M., 2013. An implantable micropump prototype for focal drug delivery. Proc. MEMEA 278–281.

Khoo, M., Liu, C., 2000. A novel micromachined magnetic membrane microfluid pump. 22nd Annual International Conference of the IEEE Engineering in Medicine and Biology Society, Vol. 3. pp. 2394–2397.

Krutzch, W.C., Cooper, P., 2001. Introduction: classification and selection of pumps. In: Karassik, I.J. , et al., (Eds.), Pump Handbook. McGraw-Hill, New York, NY.

Laser, D.J., Santiago, J.G., 2004. A review of micropumps. Micromech. Microeng. 14, R35–R64.

Saati, S., et al., 2009. Mini drug pump for opthalmic use. Trans. Am. Opthalmol. Soc. 107 (December), 60–70.

Santini Jr, J.T., Cima, M.J., Langer, R., 1999. A controlled-release microchip. Nature 397, 335–338.

Simon, H., Sven, S., Stephan, M., Roland, Z., 2012. Osmotic micropumps for drug delivery. Adv. Drug Deliv. Rev. 64, 1617–1627.

Wireless Endoscopy Capsules

15.1 INTRODUCTION

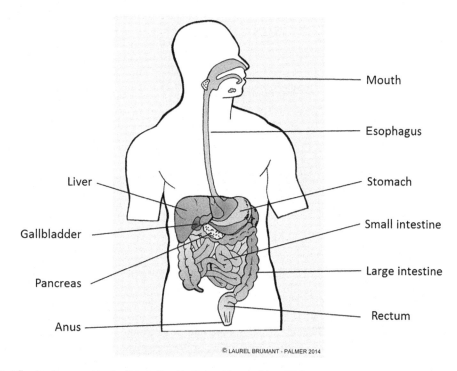

■ **Figure 15.1** Different sections comprising the GI tract. *(Copyright © Laurel Brumant-Palmer 2014.)*

The digestive system breaks food down into energy that is required by cells in the body to maintain life. The digestive system starts with the mouth, where food is chewed and mixed with enzymes in saliva before being swallowed. The food then passes down the esophagus to the stomach, where acids and enzymes break it down into a liquid or paste consistency. The food then passes into the lower or small intestine, which is also known as

Implantable Electronic Medical Devices. **DOI: http://dx.doi.org/10.1016/B978-0-12-416556-4.00017-6**

the small bowel and consists of the duodenum, jejunum, and ileum, which contain enzymes secreted from the pancreas and bile secreted from the gall-bladder to further break down the fat, protein, and carbohydrates in the food, which are then absorbed through the walls of the small intestine into the bloodstream. Muscular peristaltic contractions then push undigested food and waste products from the digestion process into the large intestine, also known as the colon, where any remaining nutrients and water are absorbed, leaving a solid waste product that is pushed through to the rectum, where it leaves the body naturally through the anus (Figure 15.1).

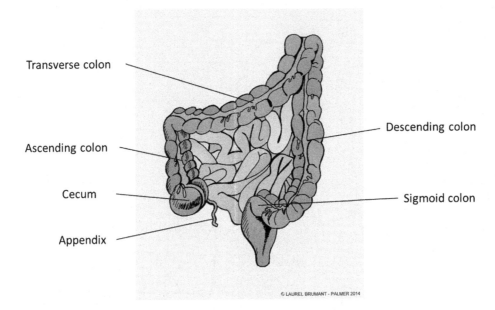

© LAUREL BRUMANT - PALMER 2014

■ **Figure 15.2** Large bowel and the different parts of the colon.

Figure 15.2 shows the start of the colon at the end of the small intestine at the cecum and consists of the ascending colon, which ascends on the right side of the abdomen; the transverse colon which goes across the abdomen; the descending colon on the left side of the abdomen; and the curved sigmoid colon which joins to the rectum. The colon and rectum are also collectively known as the large bowel.

There are many digestive conditions and disorders of the gastrointestinal (GI) tract such as obscure GI bleeding (OGIB), Crohn's disease, celiac disease, cancer of the small intestine and colorectal cancer. Many of these disorders, if detected early, can be treated. One such detection technique is flexible sigmoidoscopy, whereby a thin tube attached to an optical fiber light source and a video camera is inserted into the rectum and colon. The flexibility of the tube allows the camera to enter the sigmoid colon and the descending colon, in order to examine the inner lining for any ulcers or growths known as polyps that can subsequently be biopsied or removed. However, the camera cannot be advanced to examine the transverse and ascending sections of the colon. Air must also be pumped in to distend the colon for greater visibility of the GI tract. Pain medication is normally administered to relieve any discomfort.

Other investigative techniques include magnetic resonance imaging (MRI) and CT colonography, also known as virtual colonoscopy. CT colonoscopy uses computerised tomography (CT) technology where X-rays are used to build up a three-dimensional picture of the large bowel. An alternative noninvasive technique known as capsule endoscopy can also be used to visualize the GI tract. The procedure involves patients ingesting a wireless endoscopy capsule similar in size and shape to that of a vitamin pill which contains a miniature camera to capture images of the GI tract as the capsule naturally moves through the GI tract to the rectum, where it passes out of the body naturally from the anus.

Wireless endoscopy capsules usually incorporate white LEDs arranged around a central lens and a viewing window so that light transmitted from the LEDs illuminate sections of the wall lining of the GI tract. The light reflected back from the walls of the GI tract is directed through an optical lens to focus the image onto the semiconductor camera, which is either a charge-coupled device (CCD) or complementary metal oxide semiconductor (CMOS) image sensor that processes the light received and generates video data for data transmission to external receivers or for data storage in the onboard capsule memory.

15.2 PILLCAM® CAPSULE ENDOSCOPY BY COVIDIEN GI SOLUTIONS

There are four types of wireless capsule endoscopy systems manufactured by Covidien GI Solutions; the PillCam® SB to visualize the small bowel, the PillCam® COLON to visualize the colon, and the PillCam® ESO to visualize the esophagus. These capsules actively record and wirelessly transmit images from the GI tract to an external recorder worn by the patient. The PillCam® patency capsule is a passive device used to perform a patency test of the GI tract.

The procedure starts with the patient ingesting an endoscopy capsule. As the capsule passes through the GI tract, an LED light source flashes and images of the GI tract are captured by the onboard camera(s). Data captured from the capsule during the procedure is transmitted to an external recorder worn by the patient. Following the procedure, data from the recorder is subsequently downloaded for diagnostic review.

The three active wireless endoscopy capsules are similar in that they incorporate four white LEDs arranged around a central lens and viewing window so that light transmitted from the LEDs illuminate wall sections of the GI tract. The light reflected back from the walls of the GI tract is directed through an optical lens to focus the image onto a CMOS image sensor that processes the light received and generates video data. Radiofrequency (RF) circuits and an RF antenna, wirelessly transmit the data at a frequency of 434.1 MHz to an external data recorder. The PillCam® capsules are powered by silver oxide batteries encapsulated in a pill-like structure.

Captured images can be viewed as a continuous video stream using the proprietary RAPID® software or by using the handheld RAPID® Real-time viewer, which enables real-time continuous video as the capsules pass through the GI tract. Optimal light illumination is automatically controlled to provide an illumination field depth of 30 mm.

15.2.1 PillCam® SB 3

■ **Figure 15.3** PillCam® SB 3. *(Copyright © Covidien GI Solutions. Reprinted with permission.)*

The PillCam® SB 3 shown in Figure 15.3 is used to visualize the small bowel to detect and monitor lesions, bleeding, Crohn's disease and even iron deficiency anemia in the small bowel. The PillCam® SB 3 incorporates one video camera and wirelessly transmits at least two color images per second, resulting in more than 50,000 images for one pass through the GI tract. Using adaptive frame technology, the captured frame rate is automatically increased from two to six frames per second when the capsule is sensed to be moving more quickly through the GI tract. The viewing angle is 156°, covering an area of 1100 mm². The PillCam® SB 3 measures 26.2 mm by 11.4 mm and weighs 3 g.

15.2.2 **PillCam® COLON 2**

■ **Figure 15.4** PillCam® COLON 2. *(Copyright © Covidien GI Solutions. Reprinted with permission.)*

The PillCam® COLON 2 shown in Figure 15.4 provides a direct view of the entire colon and can be used to noninvasively complete a colon examination in patients who have undergone an incomplete colonoscopy. The PillCam® COLON 2 has two video cameras, one at each end and incorporates bidirectional wireless communication, enabling images to be captured at either 4 or 35 frames per second, resulting in approximately 50,000–100,000 images. PillCam® COLON 2 has a field of view of 172° and measures 31.5 mm by 11.6 mm, weighs 2.9 g and can operate for approximately 10 hours.

15.2.3 **PillCam® ESO 2**

■ **Figure 15.5** PillCam® ESO 2. *(Copyright © Covidien GI Solutions. Reprinted with permission.)*

The PillCam® ESO 2 shown in Figure 15.5 is used to examine the esophagus and to enable physicians to monitor and detect abnormalities, including esophageal varices. The PillCam® ESO 2 incorporates two color video cameras, one at each end and transmits 18 color images per second. The PillCam® ESO 2 has a field of view of 169°, measures 26.4 mm by 11.4 mm and weighs 2.9 g. The complete examination takes less than 30 min. The next generation PillCam® ESO 3, also contains two video cameras but transmits 35 colour images per second and has a field of view of 172°. The PillCam® ESO 3 is slightly longer than the ESO 2 capsule, measures 31.5 mm by 11.6 mm and weighs 2.9 g. The ESO 3 has been developed but is not currently available for commercial use.

15.2.4 **PillCam® patency**

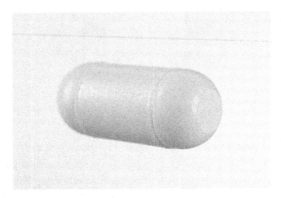

■ **Figure 15.6** PillCam® patency Capsule. *(Copyright © Covidien GI Solutions. Reprinted with permission.)*

In cases where there may be known or suspected obstructions or strictures of the gastrointestinal tract, a PillCam® Patency capsule, shown in Figure 15.6, is used to verify adequate patency of capsule endoscopy for patients. The PillCam® Patency capsule is the same size as the other endoscopy capsules and contains a barium lactose mixture and a miniature RFID tag, such that its progress through the GI tract can be detected by X-rays or by using the proprietary Patency capsule radio scanner. The capsule starts to dissolve 30 h after ingestion and if not detected, it is assumed that the capsule has passed through the GI tract and has been evacuated as waste. Detection of the RFID tag can also be detected in the stool to confirm that the capsule is not still present in the GI tract.

15.2.5 **PillCam® Sensor Belt and Recorder**

The PillCam® Sensor Belt (Figure 15.7) is worn around the waist and contains sensors that receive the transmitted video data signals from the endoscopy capsules that are subsequently processed by the DataRecorder 3, that is attached to the sensor belt to display real-time images of the GI tract on its color LED display. If the capsules are moving too quickly through the GI tract, the DataRecorder 3 can transmit control signals to the capsules to adaptively increase the frame rate from 2 to 6 frames per second. Following the

■ **Figure 15.7** PillCam® Sensor Belt and Data Recorder 3. *(Copyright © Covidien GI Solutions. Reprinted with permission.)*

procedure, the data is downloaded from the data recorder to a PC via proprietary RAPID® software for diagnostic review. Figure 15.8 shows a captured still image of the villi in the GI tract from a PillCam® SB 3 procedure.

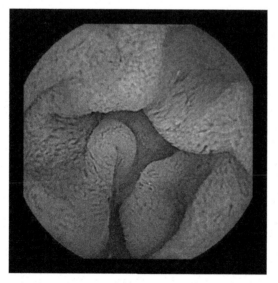

■ **Figure 15.8** Captured still image of the villi in the GI tract from a PillCam® SB 3 procedure. *(Copyright © Covidien GI Solutions. Reprinted with permission.)*

15.3 **SAYAKA ENDOSCOPE CAPSULE BY RF SYSTEM LAB**

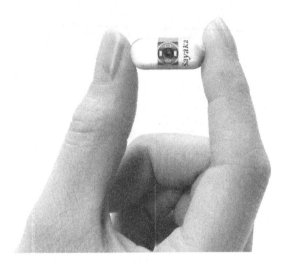

■ **Figure 15.9** Sayaka EndoScope Capsule. *(Copyright © Japanese RF System Lab. Reprinted with permission.)*

The Sayaka EndoScope Capsule (Figure 15.9) contains white LEDs and a spinning camera that faces out from the side of the capsule, such that it captures a 360° image of the walls of the intestine alongside the capsule as it passes through the GI tract. The capsule is 23 mm long and has an outer diameter of 9 mm.

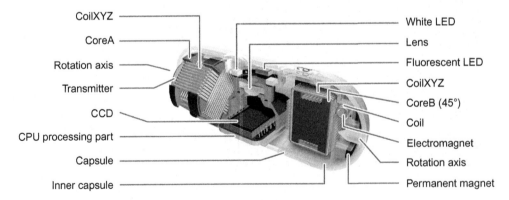

■ **Figure 15.10** Components of the Sayaka EndoScope Capsule. *(Copyright © Japanese RF System Lab. Reprinted with permission.)*

The Sayaka EndoScope Capsule contains an inner capsule (as shown in Figure 15.10) that incorporates an electromagnet and a camera attached to the same axis of a permanent magnet. Applying a current to the coils of the electromagnet, causes a permanent magnet attached to the inner capsule and the camera, to rotate around its axis. The camera is a CCD image sensor that takes 12 s to complete one rotation to provide 30° coverage of the GI lining every second. This is sufficient time for repeated close-up images considering that it typically takes the capsule 2 min to advance 5 cm in the intestine. The CCD image sensor has a resolution of 2 MB/mm^2 and captures 30 two-megapixel images per second. With an 8 h transit time through the GI tract, up to 870,000 images can be captured. The capsule is powered by the energy received from inductive coupling from charging coils in a vest worn by the patient. As the capsule passes through the GI tract, captured image data is wirelessly transmitted continuously to an antenna array in the vest. The data received is stored on a standard SD memory card for later retrieval and video playback on a PC platform.

The Sayaka EndoScope Capsule is currently still under development.

15.4 **MIROCAM® CAPSULE ENDOSCOPE SYSTEM FROM INTROMEDIC CO.**

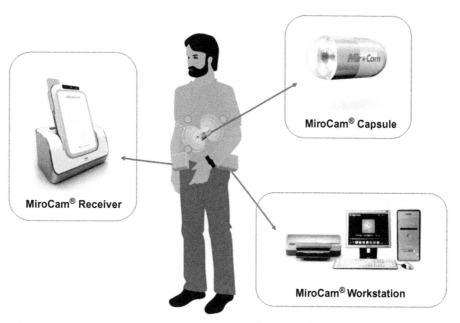

■ Figure 15.11 MiroCam® Capsule Endoscope System. *(Copyright © IntroMedic Co. Reprinted with permission.)*

The MiroCam® Endoscopy Capsule System shown in Figure 15.11 is used to investigate the small bowel by ingestion of MiroCam® endoscopy capsules that take real-time images of the lining of the GI tract as the capsule slowly moves through the small intestine. The captured image data is transmitted through the body using human body communication, patented E-Field propagation technology whereby the endoscopy capsule generates an electrical signal that is detected by a number of standard ECG electrodes placed on the patient's body. The MiroCam® Receiver, worn on the belt around the waist, compares and selects the best-quality signals from a pair of electrode channels before being processed by the MiroView® RTV proprietary software platform that enables captured images to be viewed in real time.

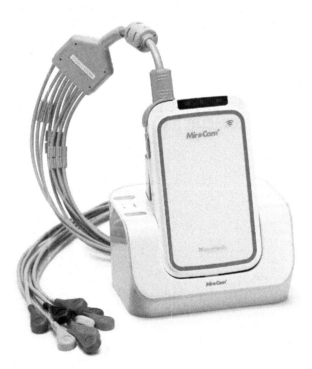

■ **Figure 15.12** MiroCam® wireless data receiver MR1100 with electrode connecting leads and charging station. *(Copyright © IntroMedic Co. Reprinted with permission.)*

There are two data receivers, the MiroCam® Receiver MR1000 which can transfer real-time image data to notebooks or laptops via USB, and the MiroCam® Receiver MR1100, which can transfer real-time image data to iPad® and iPhone® devices via Wi-Fi. The MiroCam® MR1000 is now being superseded by the MR1100 (shown in Figure 15.12) that has a higher data transmission rate and longer battery life. The MiroView™ Network System also integrates with hospital data networks to allow remote access to real-time image data.

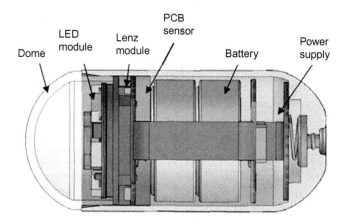

■ **Figure 15.13** Components of a MiroCam® endoscopy capsule. *(Copyright © IntroMedic Co. Reprinted with permission.)*

Figure 15.13 shows the components of a MiroCam® MC1000 capsule, which incorporates six white LEDs that flash to illuminate the walls of the GI tract. The reflected image is then focused through an optical dome onto the CCD image sensor which has a 320×320 pixel image resolution and captures 3 images (frames) per second, resulting in a minimum of 118,800 captured images. The surface of the body is gold plated to enhance signal transmission and the capsule is powered by two silver oxide batteries.

■ **Figure 15.14** MiroCam® endoscope capsules: (A) MC1000-W, (B) MC1000-WG, and (C) MC1000-WM. *(Copyright © IntroMedic Co. Reprinted with permission.)*

Figure 15.14 shows the three types of MiroCam® endoscopy capsules; MiroCam® MC1000-W, MiroCam® MC1000 WG and MiroCam® MC1000 WM. The MiroCam® MC1000-W capsule (Figure 15.14A) uses six white LEDs with a field of view of 170° and a depth of view of 7−20 mm. Captured images are transmitted at 3 frames per second with an expected operating time of 12 h. The capsule measures 10.8 mm × 24.5 mm and weighs 3.5 g.

The MiroCam® MC1000-WG capsule (Figure 15.14B) is similar to the MC1000-W in that it uses six white LEDs, has a field of view of 170°, and has a depth of view of 7−20 mm. The image frame rate is the same at 3 frames per second and the expected operating time is 12 h. The capsule also measures 10.8 mm × 24.5 mm and weighs 3.5 g. However, the MC1000-WG incorporates a mercury-free battery.

The MiroCam® MC1000-WM capsule (Figure 15.14C) also uses six white LEDs with a field of view of 170° and a depth of view of 7−20 mm. The capsule transmits 3 frames per second and has an expected operating time of 12 h. However, rather than rely on the natural peristalsis of the GI tract to move the capsule, the capsule can be magnetically steered or navigated by using the external magnetic MiroCam® Navi controller (shown in Figure 15.15).

■ **Figure 15.15** MiroCam® Navi external controller for the magnetically steered MiroCam MC1000-WM endoscopy capsule. *(Copyright © IntroMedic Co. Reprinted with permission.)*

Inside the capsule, two cylindrical magnets maintain a constant magnetic field aligned along its axis of symmetry. When an external magnetic field is applied, the axis of symmetry aligns itself perpendicular to the external magnetic field which is itself asymmetrical in order to magnetically steer the capsule. The MiroCam® MC1000-WM capsule measures 10.8 mm × 25.5 mm and weighs 4.7 g.

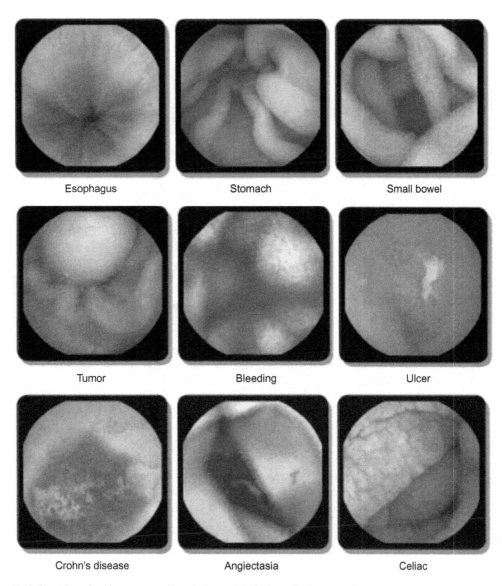

Esophagus   Stomach   Small bowel

Tumor   Bleeding   Ulcer

Crohn's disease   Angiectasia   Celiac

■ **Figure 15.16** Captured capsule endoscope images. *(Copyright © IntroMedic Co. Reprinted with permission.)*

Figure 15.16 shows images captured by the MiroCam® Capsule Endoscope System showing sections of the GI tract and digestive disorders and diseases.

15.5 **CAPSOCAM**® **CAPSULE ENDOSCOPE SV-2 FROM CAPSOVISION**

■ **Figure 15.17** CapsoCam® SV-2 capsule endoscope. *(Copyright © CapsoVision. Reprinted with permission.)*

The CapsoCam® SV-2 capsule endoscope shown in Figure 15.17 incorporates 16 white LEDs that illuminate the wall of the GI tract and four ultracompact CMOS image-sensor cameras with a 90° field of view that face outward from the side of the capsule to give a 360° panoramic view of the wall lining of the GI tract. The light intensity of the LEDs is automatically controlled to provide an optimal 20 mm illumination field depth. The CapsoCam® SV-2 endoscope capsule is 31 mm in length, has a diameter of 11.3 mm and weighs 4 g.

Images are captured during the first 2 h after ingestion at a frame rate of 20 frames per second (5 frames per second for each camera), followed thereafter by a frame rate of 12 frames per second (3 frames per second for each camera). Smart Motion Sense Technology™ ensures that the cameras are activated only when the capsule is moving. Differences between previous and current captured images are monitored to reduce redundant replicated images and to place the capsule in standby mode.

When stationary, battery power is conserved such that the batteries can operate for at least 15 h.

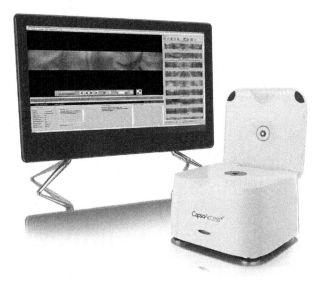

■ **Figure 15.18** CapsoAccess® capsule data access system and CapsoView® diagnostic imaging software. *(Copyright © CapsoVision. Reprinted with permission.)*

Captured image data is stored in the onboard EEPROM flash memory for later retrieval and diagnostic analysis using the CapsoAccess® capsule data access system (Figure 15.18). The capsule is placed inside the CapsoAccess® system which uses a bidirectional optical link to transfer the data stored within the onboard capsule memory via the USB interface to a PC. The captured image data is analyzed and reviewed using the diagnostic proprietary CapsoView® software. The captured images from the four cameras can be viewed in normal mode or in stack mode. In normal mode, the four concurrent camera images are aligned together in a single row to give a 360° panoramic view. In stack mode, the four camera images are arranged over two rows to give a smaller aspect ratio in which to view the images. The images can also be played back in continuous video mode. Figure 15.19 shows panoramic views of various GI tract disorders that were captured using the CapsoCam® SV-1 capsule endoscope.

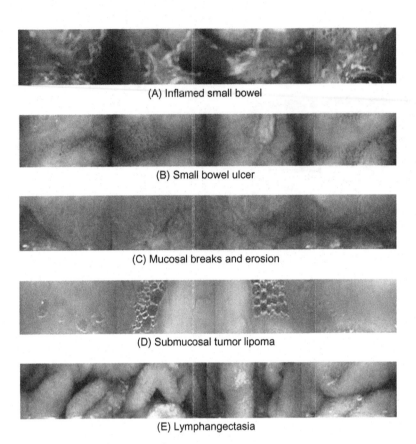

(A) Inflamed small bowel

(B) Small bowel ulcer

(C) Mucosal breaks and erosion

(D) Submucosal tumor lipoma

(E) Lymphangectasia

■ **Figure 15.19** Images of the GI tract captured by the CapsoCam® SV-1 Capsule Endoscope: (A) inflamed small bowel, (B) small bowel ulcer, (C) mucosal breaks and erosion, (D) submucosal tumor lipoma, and (E) lymphangectasia. *(Copyright © CapsoVision. Reprinted with permission.)*

15.6 **ENDOCAPSULE SYSTEM EC-S10 BY OLYMPUS, INC.**

■ **Figure 15.20** ENDOCAPSULE 10. *(Copyright © Olympus, Inc. Reprinted with permission.)*

The ENDOCAPSULE (Figure 15.20) is used to diagnose digestive disorders in the small intestine and incorporates six white LEDs with automatic brightness control to maintain optimal illumination of the wall of the GI tract to a field depth of 20 mm with a field of view angle of 160°. The ENDOCAPSULE incorporates a CMOS sensor that captures two images per second, resulting in approximately 60,000 images that can be viewed in real-time using the Smart Recorder. The ENDOCAPSULE is powered by an internal battery that lasts up to 12 h. The whole capsule measures 11 mm × 26 mm and weighs less than 4 g.

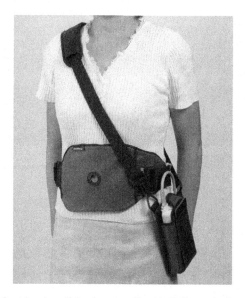

■ **Figure 15.21** Smart Recorder and belt-style antenna. *(Copyright © Olympus, Inc. Reprinted with permission.)*

The Smart Recorder is a handheld rechargeable unit that is worn in a shoulder strap pouch and connects to a belt-style antenna that is worn by the user (Figure 15.21). The recorder displays real-time images transmitted by the ENDOCAPSULE on a color LCD display and monitors its functionality and progress through the GI tract. Proprietary software enables three-dimensional tracking of the ENDOCAPSULE as it passes through the GI tract. Figure 15.22 shows GI tract images captured by the ENDOCAPSULE.

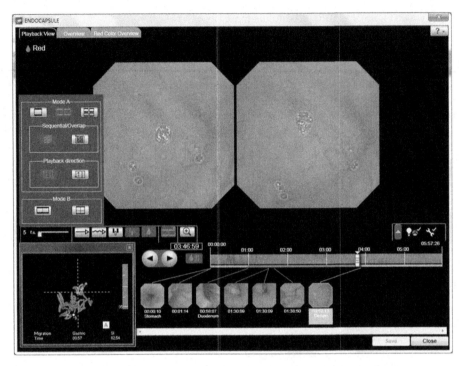

■ **Figure 15.22** GI tract images captured by the ENDOCAPSULE 10. *(Copyright © Olympus, Inc. Reprinted with permission.)*

15.7 **OMOM CAPSULE ENDOSCOPE SYSTEM BY CHONGQING JINSHAN SCIENCE & TECHNOLOGY (GROUP) CO., LTD**

The OMOM system is used to provide real-time viewing and diagnosis of disorders of the small bowel. The system shown in Figure 15.23 consists of a wireless disposable endoscope capsule, a vest worn by the user containing 14 antenna elements, a data recorder, and image software to run on a PC. The captured data is wirelessly transmitted via the vest antenna elements to the data recorder and downloaded to the PC using a USB or wireless connection for subsequent viewing and diagnostic analysis using the proprietary image software. During the capture phase, the captured images give a real-time view and estimate of the current position of the capsule.

OMOM Capsule Endoscopic System

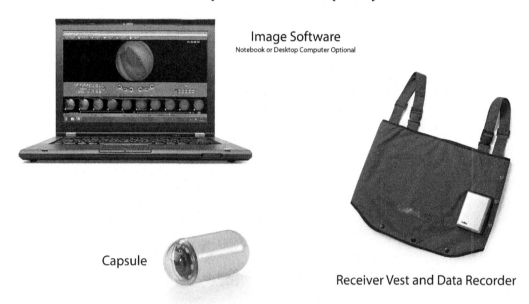

Image Software
Notebook or Desktop Computer Optional

Capsule

Receiver Vest and Data Recorder

■ **Figure 15.23** OMOM capsule endoscopy system. *(Copyright © Chongqing Jinshan Science & Technology (Group) Co., Ltd. Reprinted with permission.)*

The OMOM endoscope capsule shown in Figure 15.24 uses white LEDs and a CCD image sensor to illuminate and view sections of the GI tract with a field of view of 157° and a resolution of 0.1 mm. The capsule incorporates bidirectional wireless communication to adjust the LED illumination level and to conserve battery power by deactivating the capsule (sleep mode) when passing through the esophagus and stomach before reaching the small intestine. The frame rate can also be switched between 0.5, 1 and 2 frames per second to conserve power. Only the higher frame rate is used to view diseased sections of the GI tract with a higher resolution.

■ **Figure 15.24** OMOM capsule endoscopy. *(Copyright © Chongqing Jinshan Science & Technology (Group) Co., Ltd. Reprinted with permission.)*

The battery operating time is approximately 8 h when using a frame rate of two frames per second. The OMOM measures 27.9 mm × 13 mm and weighs less than 6 g. Figure 15.25 shows captured images of the GI tract using the OMOM endoscope capsule.

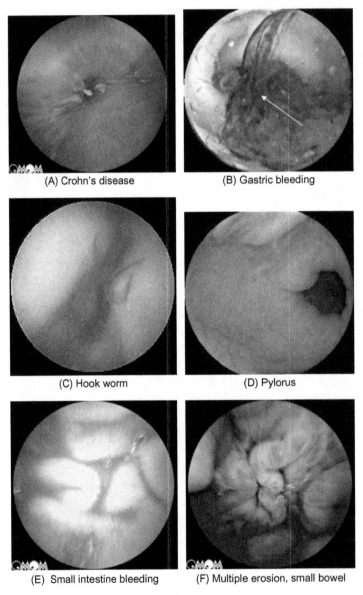

(A) Crohn's disease   (B) Gastric bleeding

(C) Hook worm   (D) Pylorus

(E) Small intestine bleeding   (F) Multiple erosion, small bowel

■ **Figure 15.25** Captured images using the OMOM capsule endoscope: (A) Crohn's disease, (B) gastric bleeding, (C) hook worm, (D) pylorus, (E) small intestine bleeding, and (F) multiple erosions, small bowel. *(Copyright © Chongqing Jinshan Science & Technology (Group) Co., Ltd. Reprinted with permission.)*

Index

Note: Page numbers followed by "*f*" and "*t*" refer to figures and tables, respectively.